T0305593

RECOGNISING AND RESPONDING TO ANIMAL EMOTION IN A SHARED WORLD

How is it that depending on the setting, the same cat can be perceived as a homeless annoyance, a potential research subject or a thinking and feeling family member? The answer is bound up in our perception of non-human animals' capacity to experience emotions, and this book draws on contemporary evidence-based research, observations, interviews and anecdotal case scenarios to explore the growing knowledge base around animal emotion. Acknowledging that animals can experience feelings directly affects the way that they are perceived and treated in many settings, and the author explores the implications when humans apply—or ignore—this knowledge selectively between species and within species.

This information is presented within the unique context of a proposed hierarchy of perceived non-human animals' emotional abilities (often based on human interpretation of the animal's emotional capacity), with examples of how this manifests at an emotional, spiritual and moral level. Implications for specific groups living with, caring for or working with non-human animals are examined, making the book of particular interest to those working, studying or researching in the veterinary professions; animal ethics, law and welfare; and zoology, biology and animal science.

This book will also be fascinating reading for anyone interested in simply learning more about the animals with whom we share this planet. For some readers, it will validate the reciprocal emotional bond they feel for living creatures. For others, it will raise questions about the moral treatment of sentient non-human beings, breaking down the human protective barrier of cognitive dissonance and activating a cycle of change.

RECOGNISING AND RESPONDING TO ANIMAL EMOTION IN A SHARED WORLD

Vicki Hutton

CRC Press

Taylor & Francis Group

Boca Raton London New York

CRC Press is an imprint of the
Taylor & Francis Group, an **informa** business

Cover credit: © Ana Paula Grimaldi

First edition published 2024
by CRC Press
6000 Broken Sound Parkway NW, Suite 300, Boca Raton, FL 33487-2742

and by CRC Press
4 Park Square, Milton Park, Abingdon, Oxon, OX14 4RN

CRC Press is an imprint of Taylor & Francis Group, LLC

©2024 Vicki Hutton

ISBN: 9781032287799 (hbk)
ISBN: 9781032287782 (pbk)
ISBN: 9781003298489 (ebk)

DOI: 10.1201/9781003298489

Typeset in Bembo
by codeMantra

For my parents who taught me that every animal counts; for my children who continue our family culture of care for all animals; and for the many animals who have shown me how to love and respect their world.

CONTENTS

ABOUT THE AUTHOR

Vicki Hutton is passionate about animals and their welfare, having lived with, worked with, observed and written about animals for many years. After starting her career in the health and community sectors as a mental health therapist, crisis counsellor and group facilitator, Vicki moved into education and research while completing her doctorate in Applied Psychology. Vicki has published widely in the human–animal relationship field, as well as the areas of mental health, stigma and discrimination and social cohesion. She currently lives with an assortment of animals who all contribute in their unique ways to her writing.

ACKNOWLEDGEMENTS

This book arose out of a personal desire to learn more about animal emotion, but it could not have progressed without input from the many animals and animal supporters who shared their stories and photographs with me: Animals Asia, Buttercups Sanctuary for Goats, Edgar's Mission, Ingrid's Haven for Cats, Lefty's Place Farm Sanctuary, Liberty Foundation Australia, Oscar's Law and the individuals who just wanted to talk about animals.

Thank you to my daughter Rebecca who accompanied me on a number of expeditions to see the many animals mentioned in this book and who generously let me include some of the photographs she took along the way.

And thank you to the animals in this book who allowed me into their lives to get some understanding of their feelings and their stories, including my current animal family who sat with me for endless hours on this shared journey into the emotional world of animals.

INTRODUCTION

It would start with a cardboard box carefully wedged on the back seat of my father's small Volkswagen beetle car. With mounting excitement, we would listen to the sounds emanating from the box and try to guess who was coming to stay the night—and in some cases, to stay forever. My mother would shake her head in resignation, but even she could not resist holding her breath in anticipation as my father sheepishly revealed the latest stray cat or cats that he had saved from potential death.

My early childhood memories are populated by the many rescue cats that came and went from our household. Mother cats and their kittens, pregnant queens and rambunctious tomcats arrived in the Volkswagen rescue vehicle to be socialised, rehabilitated, desexed and rehomed. Our house became a proxy no-kill animal sanctuary in the days when euthanasia was inevitable for those cats who reached their time limit in the over-crowded formal shelters.

This was the early 1960s when my father had joined a newly established university in Australia and the first *Australian Code for the Care and Use of Animals for Scientific Purposes* (1969) was still some years away. Stray cats wandering around the university grounds were at risk of vanishing into the zoology department laboratories for future experimentation. Decades before it became widely accepted and proven knowledge, my father knew intuitively that these cats had feelings. They could experience fear and pain in the university laboratories and comfort and joy in the safety of our family home and their future forever homes. As a child I too had no doubt that these cats—my adoptive brothers and sisters—and by extension all animals, were individuals capable of experiencing a full and meaningful emotional life and therefore to be fiercely protected from suffering.

My childhood black and white thinking would be tested many years later when my mother was diagnosed with Parkinson's disease, a progressive degenerative neurological condition. From Nobel Prize winner Arvid Carlsson, who used rabbits and mice to discover the role of the neurotransmitter Dopamine in this disorder, to studies replicating the distressing symptoms in monkeys so as to identify treatment methods, animal research has defined the medical approach to Parkinson's disease (Understanding Animal Research, 2023). With an ageing global population, animal research shows no sign of abating in the search to constantly improve treatments and ultimately find a cure for this debilitating condition.

DOI: 10.1201/9781003298489-1

Life was simpler as a child when the unwavering belief that animals were as important as humans had not yet been tested. The bitter-sweet relief that our family felt at every scientific advance in the treatment of Parkinson's disease was juxtaposed against the niggling knowledge of how this came about. I became stuck in what philosopher Strachan Donnelley (1989) described as "the troubled middle" of ethical obligations to animals, with no way out. The resultant mental conflict turned my childhood black and white thinking into a murky grey that would never subside as my passion and life's work moved further and further into the emotional lives of vertebrate and invertebrate animals.

The use of animals in invasive or non-invasive, but ultimately physically or psychologically distressing, biomedical research remains an area of contention, with justification often grounded in utilitarian ethics (Gruen, 2021). To conduct research on involuntary animal subjects requires a belief that the pain, suffering and death experienced by these mice, rats, rabbits, cats, dogs, fish or other creatures is outweighed by the hoped-for benefits to humanity. While seeming to make sense when a loved one is suffering from an incurable disease and in need of evermore sophisticated medications, the commodification of animals for human benefit does not stop there. Directly or indirectly, humanity's potentially irreversible impact on the planet is encroaching on every aspect of animals' lives, whether in research and animal testing laboratories, factory farms, illegal trafficking or degradation of the natural environment.

However, there remains a persistent and escalating movement to respect the lives and feelings of all animals as they try to survive in this human-dominated world. In 1985, animal rights proponent and philosopher Tom Regan argued that we should recognise all creatures as having rights and with mental lives that include perceptions, desires, beliefs, memories, intentions and a sense of future. The possibility that non-human animals have emotional and mental lives has intrigued humans for centuries. Eighteenth-century philosopher Jeremy Bentham asked a simple question that would become the mantra of future animal protection movements and shift the focus from an animal's capacity to reason, considered the gold standard to differentiate humans from animals, to the animal's capacity for suffering

> Is it the faculty of reason or perhaps the faculty of discourse? But a full-grown horse or dog, is beyond comparison a more rational, as well as a more conversable animal, than an infant of a day or a week or even a month, old....The question is not, Can they *reason*? Nor, Can they *talk*? But, Can they *suffer*?
>
> (Bentham, 1789, chapter xvii, footnote)

Evaluating the consequences of animal treatment in terms of pleasure and pain made sense for nineteenth-century animal abuse issues such as cock fighting and fox hunting but Bentham still privileged humans over animals when it came to research. Animal research, he believed, was acceptable when beneficial to humanity and with a reasonable prospect of success (Boralevi, 1984). With qualifiers such as this, it was clear that turning to the philosophers could provide no comfort or solution to the troubled middle in which I was stuck. Instead, positioning animals as capable of suffering but negating this suffering by virtue of their inferior status when compared to humans created a hierarchy that would eventually permeate most human–animal interactions.

Ongoing scientific research in evolutionary biology, cognitive ethology and neuroscience provides convincing evidence that many animals share neuroanatomical structures and neurochemical pathways fundamental to core emotions. It is many years since American biologist, ethologist, behavioural ecologist and writer Marc Bekoff commented that "the more we study animals and the more we learn about 'them' and 'us', the more we discover there is no real dichotomy or nonnegotiable gap between animals and humans, because humans are, of course, animals" (2005, p. 12). This sentiment was formally ratified in 2012 when a prominent group of scientists signed the *Cambridge Declaration on Consciousness*, which affirmed that humans are not unique in possessing parts of the brain complex enough to support conscious experiences. The Declaration identified that a significant number of nonhuman animals, including all mammals and birds and some other animals such as octopuses, could feel what was happening to them and experience positive and negative states such as fear, play, anger, irritation, love, sadness and grief.

Although difficult to assess and validate the extent and form of these conscious experiences in animals who could not articulate their feelings in a human-like manner, escalating scientific neurological data related to pain and pleasure had serious welfare implications. These implications were not confined to the mammals, birds, fish, reptiles and amphibians that comprise the more familiar five groups of animals with backbones, the vertebrates. The Center for Biological Diversity (2023) estimates that invertebrate animals without an internal backbone, including the octopuses mentioned in the Declaration, account for around 97% of animal species on Earth. Many of these invertebrate animals are informally grouped under the label of insects, of which there are an estimated 1–10 quintillion, or 200 million insects for every human on Earth (Sverdrup-Thygeson, 2019). The flow-on ethical implications of potential suffering in even a small proportion of these invertebrates could be immense.

There is a long history of using mammals in research, premised on the assumption that their physiology is similar enough to humans to obtain valid results. While the similar physiology aspect supports Bekoff's notion that there is no real them and us, the procedures that these animals are subject to reinforce the stark divide between human animals and nonhuman animals. Many of the drugs for Parkinson's disease that temporarily eased the tremors, shuffling gait and rigidity plaguing my mother's daily life related to the neurotransmitter, Dopamine, located in several regions of the brain including the substantia nigra. As well as being related to movement, Dopamine influences mood and feelings of reward and motivation. To be meaningful research models, the mice, rats and rabbits must share this neurotransmitter and its effects on brain processes but remain subject to the anthropocentric division that positions one as the test subject and one as the beneficiary.

Many psychiatric medications such as mood stabilisers are tested on animals, a further reminder that animals have emotions underpinned by the same mechanisms as humans and are responsive to the same drug manipulations. As animal behaviourist Temple Grandin (2018) succinctly commented: "If the nervous system of a rat was totally different from that of humans, this research would be totally useless" (p. 16). It is therefore not surprising that veterinarians often prescribe human psychiatric medications "off label" or "extra label" to treat certain behavioural conditions in dogs, cats and birds. Prozac (fluoxetine), an SSRI antidepressant, can be used to treat separation anxiety in dogs while Alprazolam (Xanax) is a sedative used to treat anxiety and phobias in cats and dogs (Gollakner, 2023).

Acknowledging that some animals can experience feelings—although not necessarily in a way that humans can understand—directly affects the way that different animals are perceived and treated in many settings, not just research laboratories. Animals are eaten, worn, deprived of their freedom and natural environments, legally and illegally sold, blamed for pandemics and used as a buffer to loneliness, to name just a few. In all these scenarios, the existence and extent of their emotional capacity is perceived differently both between and within species, with a direct flow-on effect on their life experiences and wellbeing in a human-dominated world. Interrogating the basis for these at times contradictory perceptions of different species, or members of the same species in different environments, reveals the extent of human supremacy. It also reveals that perception of emotion in animals is not always driven by biological facts. In many cases, this perception is based on a social construct created and maintained by human beings as the more dominant animal.

Decades later as I sift through the photos of our many feline adoptees, I question how Princess, the tabby cat who gave birth in the back of an engineering toolshed and her four tiny kittens, Paddy, Pud, Sheba and Tiger crouching behind their protective mother, could simultaneously be perceived as homeless annoyances, potential test subjects outside of any ethical guidelines or family members with emotional capacity. The answer is simple: it depends on how humans have categorised and labelled them. In my father's case, he perceived each cat as an individual with feelings, a unique personality and the right to live in safety (Figure 0.1).

Figure 0.1 Princess sits on a comfortable cushion in the safety of her new home.

Categorising animals reinforces the social constructs surrounding their worthiness and perceived emotional capacity and makes it easier to treat them differently. Melanie Challenger (2021), researcher on the history of humanity and the natural world, commented that this "them and us" scenario was sustained by the prefix "nonhuman", which positioned animals as separate and lesser and bestowed possession and other rights on the higher human animal. For centuries, humans have sought to rank animals and their abilities. As creators of these rankings and hierarchies, humanity always ended up as the pinnacle of evolution and therefore the gold standard. The possibility that animals have emotional capacity not only presents a threat to humanity's self-appointed exceptionalism but catapults the thinker back to the troubled middle of ethical obligations and the mental discomfort of holding conflicting views or beliefs.

Human perceptions can play a large part in determining which vertebrate and invertebrate animals have sufficient emotional capacity to make them worthy of societal protection and which animals can be classified as commodities or problems, with minimal or no societal protection. How that fragile subjective framework is constructed and enforced can have long-term welfare implications for animal wellbeing. The Zoological Emotional Scale introduced in Chapter 2 and operationalised throughout this book provides a lens through which to explore the role of human perception in quantifying an animal's emotional capacity, the mechanisms fuelling those perceptions and the subsequent implications for the animal. It provides some insight into who decides which animals are more equal than others, and what can be done about this. Importantly, it offers a tool with which to explore the question—Why do I think this way?

Recognising and defining emotion is not easy, especially when expecting animals to fit human models of emotional responses. In Chapter 3, the terms consciousness, sentience and emotion are clarified so as to contextualise the subsequent chapters relating to the emotional lives of animals in family, love and loss, the spirituality and culture of animals, and the moral implications of co-existing with animals fully capable of living an emotional, spiritual and moral life if given the opportunity to do so. Animals do have feelings and voices, but they are often not heard. This book explores one way for humans to hear and respect animal voices, no matter what language they are speaking.

REFERENCES

Bekoff, M. (2005). *Animal passions and beastly virtues: Reflections on redecorating nature.* Temple University Press.

Bentham, J. (1789). *An introduction to the principles of morals and legislation.* Clarendon Press.

Boralevi, L.C. (1984). *Bentham and the oppressed.* Walter de Gruyter.

Cambridge Declaration on Consciousness. (2012, July 7). Written by Philip Low and edited by Jaak Panksepp, Diana Reiss, David Edelman, Bruno Van Swinderen, Philip Low and Christof Koch. University of Cambridge.

Center for Biological Diversity. (2023). *Invertebrates.* https://www.biologicaldiversity.org/species/invertebrates/

Challenger, M. (2021). *How to be animal. A new history of what it means to be human.* Canongate Books Ltd.

Donnelley, S. (1989). Speculative philosophy, the troubled middle, and the ethics of animal experimentation. *Hastings Center Report, 19*(2), 15–21. https://doi.org/10.2307/3563134

Gollakner, R. (2023). *Alprazolam.* VCA Animal Hospitals. https://www.biologicaldiversity.org/species/invertebrates/

Grandin, T. (2018). My reflections on understanding animal emotions for improving the life of animals in zoos. *Journal of Applied Animal Welfare Science, 21*(supp1), 12–22. https://doi.org/10.10 80/10888705.2018.1513843

Gruen, L. (2021). *Ethics and animals: An introduction.* Cambridge University Press.

Regan, T. (1985). The case for animal rights. In P. Singer (Ed.), *In defense of animals* (pp. 13–26). Basil Blackwell.

Sverdrup-Thygeson, A. (2019). *Extraordinary insects.* Mudlark.

Understanding Animal Research (2023). *Parkinson's Disease.* https://www.understandinganimalre-search.org.uk/why/human-diseases/parkinsons-disease

THE NEED FOR ORDER

Humans have a long history of trying to establish and maintain order in their world. They arrange, label, rank, predict and file to retain a semblance of feeling in control. However, unlike paper and electronic filing systems, the natural world does not always fall into neat categories, nor does it remain within the artificial constraints of human regulation. The Zoological Emotional Scale, which is discussed in more detail in the following chapter, is not presented as another structure of order. Rather, it draws attention to the anomalies embedded within a person's rationale for ranking which, if any, animals have feelings. The subjective nature of the scale is evident in the subliminal appraisal of animals in four domains—friend, foe, functional or based on fallacy—which can affect a person's perception of the animal's worthiness to be deemed an emotional being. Privileging one or more of these domains during the appraisal of emotional capacity and subsequent treatment of an animal can have far-reaching implications for the individual animal, the species and in some cases, the environment. Given the many different ways that humans interact with animals, inevitably animals from the same species may be positioned in more than one domain or move between domains, as evidenced by the university cats who shared my childhood. The short journey on the backseat of the Volkswagen beetle car shifted those fortunate few homeless cats from functional commodities whose worth lay in what their bodies could offer science to the friend domain where their emotional capacity was unquestioned and reciprocated as beloved family members.

How an animal achieves the status of friend, foe, functional or fallacy can be based on a mix of factors, including the visual appearance of the animal (especially emotive, paedomorphic features), a person's first-hand experience with the animal or representative animals and learned beliefs and attitudes. It can also depend on the mechanisms and strategies invoked to manage the cognitive dissonance, or unpleasant feelings, that may arise when trying to reconcile using and abusing animals to satisfy personal needs and wants. In 2010, anthrozoologist Hal Herzog published a provocatively titled book, *Some We Love, Some We Hate, Some We Eat: Why It's So Hard to Think Straight about Animals*, in which he opened with the statement: "The way we think about other species often defies logic" (p. 1). People employ a number of complex psychological strategies to subdue the potential mental pain of these logic-defying interactions so succinctly summed up by Herzog's book title.

Social media and mainstream media regularly confront the reader with images of factory-farmed, mass-produced pigs, cows, chickens and fish who live and die in

DOI: 10.1201/9781003298489-2

squalid, artificial environments. Grieving vocalisations when mothers are separated from newborns and self-harm and repetitive behaviours known as stereotypies provide indisputable evidence of human-inflicted suffering among animals, and yet these practices continue. Cognitive dissonance is an unpleasant feeling that can arise whenever a person holds two thoughts, or a thought and a behaviour, that contradict each other. Seeing these images and eating meat is a guaranteed recipe for cognitive dissonance and leaves a person looking for options to relieve this discomfort. Whether that is finding some justification for the continuation of these practices, or simply ignoring the evidence because it hurts, the end result is the same—continuation of what has come to be known as the "meat paradox".

The meat paradox describes a belief-action gap between professed care for animals and simultaneously indirectly harming them through eating, drinking or wearing parts of their bodies (Scott et al., 2019). Arising from a mix of socialisation, social and cultural norms, economic constraints to behavioural change or institutional constraints to individual change, it can be as hard to shift as the "troubled middle" of ethical obligations to animals in research. Strategies to quell the resultant discomfort of cognitive dissonance include a denial of personal responsibility, an attack on those who would censure a person's belief-action discrepancies or the justification of a person's actions within the broader economic environment. Dissociating from the process of meat production and its effect on the individual animal and the environment is also a viable strategy, facilitated by the physical distancing of a piece of packaged supermarket meat that bears no resemblance to the living creature from whom it originated.

Understanding how people arrive at their beliefs and assumptions of animal emotional capacity is not an exact science, but it is an important step in improving the welfare of vertebrate and invertebrate animals. As evidence accumulates to support Marc Bekoff's (2005) contention that there is "no real dichotomy or non-negotiable gap between humans and animals" (p. 12), it becomes harder to ignore the disparity in treatment between human animals and nonhuman animals. Viewing this disparity through the lens of the Zoological Emotional Scale offers some insight into the complex world of accepting or rejecting the notion of animal emotion. But first, it is important to briefly look at the historical legacy and humanity's need for order that has consolidated the legitimacy of an animal kingdom hierarchy with humans at the apex.

HIERARCHIES AND THE NEED FOR ORDER

Hierarchies in the natural world proliferate. A wolf pack maintains order through its hierarchy, reinforced by displays of dominance and submission. Wasps live in a hierarchical society where it is important to understand who is in charge. Chickens live in a stable social group where they recognise and memorise the faces of other flock members and quickly learn their place in the pecking order. Termites form large colonies based on a caste structure that allows them to create mounds several metres high. Hierarchies can offer safety, security and power, and so it is no wonder that for centuries humans have equated hierarchies in nature with stability and sought to emulate this on a grander scale.

The science of biological classification—known as taxonomy—provides a classification system designed to bring order to the complex world of nature. A hierarchical model of animal life has been embedded in Western scientific discourse since Aristotle's conception of nature as ordered on a vertical scale that extended from lifeless things to humans. Swedish botanist Carolus Linnaeus formalised biological classifications into a hierarchical system with naming protocols and arrangement of living entities based on shared characteristics. The rank of species brought together those animals that most resembled one another and which were at a stage of evolution where they were capable of interbreeding. Species, thus, consisted of the most homogeneous groups it was possible to define each clearly separated from the rest (Cain, 1993). The International Code of Zoological Nomenclature (2012) continued the concerted effort to establish and maintain order in nature by sorting out relationships to provide a logical and consistent system of nomenclature with eight categories: domain, kingdom, phylum, class, order, family, genus and species.

Inclusion of these biological classification systems seems far removed from emotions in animals, but it does help contextualise lingering disparities in the way humans perceive and treat animals. As any historian will attest, sometimes it is necessary to look backwards in order to move forwards. Centuries of hierarchical thinking that privileges humans can be hard to shift, especially when looking at the phylum category where an undeniable hierarchy based on animal complexity permeates past and present thinking. Those organisms least resembling humans physically and therefore by assumption, mentally, were positioned at the bottom, thus visually consolidating their inferiority and the lingering division of living creatures into higher and lower groupings. The bottom tier of the hierarchy is inhabited by sponges (Porifera), before moving up to coral, jellyfish and anemones (Cnidaria), tapeworms and flukes (Platyhelminthes), earthworms and leeches (Annelida), invertebrate clams, mussels and snails (Mollusca), invertebrate insects, crustaceans and arachnids (Arthropoda) before arriving at the top tier and the vertebrate Chordata, of which humans comprise only a tiny portion (Cain, 1993). Variations of hierarchical scales dating back to Charles Darwin automatically positioned humans with their unique traits and abilities at the top. This was despite growing evidence that separation by hierarchy is at the least outdated and at the worst, contributing to the maintenance of less than desirable treatment of animals. The flow-on effect of this division and categorisation was poignantly summed up by a researcher on the history of humanity and the natural world, Melanie Challenger (2021), when she wrote: "The world is now dominated by an animal that doesn't think it's an animal. And the future is being imagined by an animal that doesn't want to be an animal" (p. 1).

COGNITION, BRAINS AND CULTURE

While phylogenetic systems of classification rank animals on the basis of biological distinctions, other systems draw on cognition, intelligence and brain size to categorise and order animals on degrees of comparison—and in many cases inferiority—to human capacities. Humans have an irresistible need to compare animal intelligence to their own, as evidenced by research studies and resultant headlines that always seem slightly bemused when announcing that: "Crows are as smart as a seven-year-old child, researchers find" (2014) or "Pigs are gentle creatures…smarter than dogs and even 3-year-old children" (2023), and the unsurprising finding that "Human-like intelligence in animals is far more common than we thought" (Robson, 2021).

Many decades ago, it was acknowledged that humans can have multiple types of intelligence. Psychologist Howard Gardner (1983) proposed a theory of multiple intelligences while Robert Sternberg contended that human intelligence could not be meaningfully understood outside of its cultural context. Importantly, Sternberg and Grigorenko (2004) stressed that behaviour considered intelligent in one human culture could be perceived as unintelligent in another culture, a timely reminder that animals do not need to mimic human-like intelligence to be geniuses within their own culture and environment. While recognition and acceptance that animals have different cognitive abilities beyond the scope of human capabilities have gained traction, there is a lingering tendency to perceive them within a human perspective, thus maintaining a simplistic hierarchical approach to arranging mental capacity among species.

Brain size can be a factor when comparing intelligences among animals and humans, but a ranking based on brain size alone would see humans pushed down the scale behind elephants, whales and some dolphins. The Cephalization Index attempts to equalise disparities in brain size by measuring how big a brain is after taking into account body size. Humans are back to the top of the list with this measure, followed by dolphins, chimpanzees, monkeys, elephants and whales (Wynne & Udell, 2013). This has been further refined to consider how much of the brain comprises the neocortex, that part of the brain believed to reflect advanced intelligence, and how much brain folding into grooves and ridges is needed to fit the neocortex into the skull. Because the neocortex is associated with intelligence, and humans have so much neocortex that it will only fit into the skull with multiple folding, humans are back to the top of intelligence. The human neocortex comprises 80% of brain weight, while primates slip down the index at 50%, rodents at 30% and insectivores at 13% neocortex.

Crows might be "as smart as a seven-year-old child", but birds developed in different ways to mammals and cast doubt on the notion that relative brain size and neocortex are measures of intelligence. A bird's forebrain, although lacking the layered structure of the neocortex, is responsible for many of the functions involved in higher cognition and therefore difficult to classify within the Cephalization Index. Birds have also been described as natural "split-brains" meaning they have no structure similar to the corpus callosum that connects the two hemispheres of the human brain, raising the possibility that a bird may comprise a pair of conscious subjects (Birch et al., 2020) and negating the descriptor "bird brain" that humans use derogatively to describe a silly or stupid person. Even more intriguing is the cerebral ganglia and brachial plexus (nerve ring around the top of the arms) of the octopus, suggesting that octopuses may have up to nine conscious perspectives of the world but no neocortex. The number of nerve cells within the brain can further complicate comparisons. An individual honeybee brain contains up to a million neurons compared to around 100 billion in the human brain (Menzel, 2012). The bees' hive mind, however, contained within the bee colony's superorganism living entity status (or Bien) contains billions of neurons. The shared knowledge residing in the minds and culture of the Bien shifts intelligence away from the individual to the larger group within this bee culture. When describing the uniqueness and importance of animal culture, ecologist and author Carl Safina (2020) commented that assuming animals do not have shared knowledge because they do not have human culture and learning is nonsensical.

This again raises the question posed earlier: why are comparisons with human capabilities necessary at all other than to maintain humanity's self-appointed position at the top of the hierarchy?

ALTERNATIVE, MORE FLEXIBLE SCALES

Moving away from biologically based scales and hierarchies, sociologist and anthropologist Arnold Arluke and writer Clive Sanders (1996) proposed a sociozoological system to rank animals based on how well they were judged to play the roles expected of them in society. Seeking to understand the paradox of humans loving some animals as sentient creatures and maltreating or killing others as utilitarian objects, Arluke and Sanders pondered how it was possible for some humans to fail to examine—or even be aware of—this contradiction.

Expanding on the Sociozoological Scale in their book *Regarding Animals*, Arluke and Sanders contended that an animal's worth and position on the social ladder was largely determined by the animals themselves and their ability to know their place in human society and stay within it. Good animals willingly accepted their subordinate place in society, either by seeming to enjoy it (for example, companion animals) or complying with it because they were powerless to do much, if anything, about it (for example, farm animals and research animals).

Humans, in their role as animals who failed to believe they were animals, topped the scale, followed by the affectively or instrumentally good animals who remained obedient in their given roles. To be instrumentally useful as mechanisms for human benefit, Arluke and Sanders proposed that animals must be de-anthropomorphised into lesser beings or objects that have few thoughts and primitive, if any, emotions. To perceive them as otherwise opens the door to cognitive dissonance, something the human brain strives to overcome. At the opposite end of the Sociozoological Scale were those animals that Arluke and Sanders described as freaks who mistook their place in society, vermin who strayed from their acceptable place in society, or animals who refused to accept human power over them and therefore rejected their place in society. The freaks, vermin and rejectors embodied a narrative of disruption and harm, thus allowing humans to avoid the painful conflict of cognitive dissonance at the destruction of members of these groups. Nevertheless, Arluke and Sanders also made the point that when considering modern Western attitudes to animals, "one of the most glaring consistencies is inconsistency" (1996, p. 4), reinforcing the notion that categorisation and treatment of animals is dependent on an unstable mix of many different variables.

Professor of animal ethics and welfare James Serpell (2004) developed a model of human attitudes to animals that was based on a grid framework. Within that grid were two motivational considerations: affect, depicting people's emotional responses to animals; and utility, depicting people's perceptions of the animal's instrumental value. Tension arose when an animal was positioned on the grid with a strong positive utility value simultaneously with a strong positive affect, largely because the utilitarian treatment could result in harm or death. Sometimes, an animal's temperament alone could place them in differing dimensions of the grid, as evidenced by the beagle dog.

Picture a beagle with large expressive eyes, soft flowing ears and a gentle temperament, characteristics that position these popular companion animals on the strong affect dimension of the grid. These same temperament characteristics have seen beagles shifted to the strong utilitarian dimension as preferred dogs for laboratory experiments where they have remained for decades.

From the 1950s to the early 1980s, beagles were part of Atomic Energy Commission-funded tests at various universities and laboratories across the United States. This was to ascertain the physical symptoms and lifespan effects of radiation poisoning to help humanity prepare for potential nuclear warfare (Bolman, 2022). Whether a beagle achieved beloved family member status or was subject to radiation poison symptoms such as bone tumours, spontaneous bone fractures, tooth loss and skeletal disfigurations creates tension when trying to reconcile the same animal at opposite dimensions of the grid.

Serpell's grid, the Sociozoological Scale and others that followed rejected the simplistic linear hierarchy with higher animals having more complex abilities, self-awareness and merit. Instead, they gave voice to the potential incongruence of human perceptions and beliefs with the reality of an animal's lived experience. Gordon Burghardt, professor in psychology, ecology and evolutionary biology, and Harold Herzog (1980) used the analogy of a ball of Play-Doh that could be flattened and stretched in many directions when considering the intense, complex and paradoxical feelings that humans exhibit towards other creatures. The Play-Doh analogy is particularly relevant as it allows for a constant shifting and re-moulding of attitudes, essential if human treatment and respect for animals as unique individuals is to be interrogated and re-aligned.

Animals exist not only as a species that must be managed, protected, utilised— or abused—but also as unique individuals whose lives matter. People are quick to point out the dangers of stereotyping other humans and the flow-on effects of discrimination and violence, but this is equally applicable to animals. Perceiving an animal as an individual rather than part of a homogenous group circumvents the erroneous task of attempting to categorise simply to satisfy the human need for order within the turmoil of the natural world. Nadine Dolby (2019), founder of Animal Advocates of Greater Lafayette, commented that leaving aside anthropocentric biases could shift the paradigm to a common worlds approach where humans learn with animals rather than about them so as to address the challenges facing all living creatures.

A willingness to accept that animals, like humans, are unique individuals also acknowledges that an animal's emotional capacity will never manifest in a prescribed and simplistic human-like manner. People display a range of contradictory behavioural representations of emotions as evidenced by the experience of sadness. One person may resort to laughter and black humour while another directs anger outward or inward, and yet another is tearful and incapacitated. All can be symbolic of deep sadness, but each is unique to the individual. Psychologists and psychotherapists may be called in to help a person make sense of and regulate their emotional responses, drawing attention to the impossibility of attempting to compare human emotions to animal emotions when behavioural responses can be so diverse and mercurial even within a species that share a common language.

While neuroscience provides some insight into the biological basis of human and animal emotion, there remain some disparities in how this translates into the treatment of different animals, both within the same species and between species. Anthropologist Barbara King (2021) illustrates this distinction when contrasting cows relaxing in an animal sanctuary where they have names and unique personalities to functional cows with plastic portals in their sides at a university research facility. Understanding how humans perceive animal emotional capacity may help contextualise the contradictory treatment of cows and other animals whose quality of life is dependent on human attitudes and needs.

However, interrogating embedded attitudes and beliefs not only risks cognitive dissonance, but may also spark legal debate, especially in relation to non-mammals and less familiar animals. With growing evidence that crustaceans and cephalopods are sentient creatures who experience pain, the life and death of a lobster created a media and legal storm in some countries. Whether lobsters and crabs should be dropped live into boiling water or more humanely killed now depends on a country's animal welfare legislation and their geographic location. The following chapters seek to create a narrative within which to interrogate the animal rankings that permeate human thinking and to give voice (albeit still a human voice) to the emotional life of the vertebrate and invertebrate animals with whom we share this planet.

REFERENCES

Arluke, A. & Sanders, C.R. (1996). *Regarding animals*. Temple University Press.

Bekoff, M. (2005). *Animal passions and beastly virtue: Reflections on redecorating nature*. Temple University Press.

Birch, J., Schnell, A.K., & Clayton, N.S. (2020). Dimensions of animal consciousness. *Trends in Cognitive Sciences, 24*(10), 789–801. https://doi.org/10.1016/j.tics.2020.07.007

Bolman, B. (2022, August 22). Beagles are in the news after decades as key players in medical research. *The Washington Post*. https://www.washingtonpost.com/made-by-history/2022/08/26/beagles-are-news-after-decades-key-players-medical-research/

Burghardt, G. M., & Herzog, H. A., Jr. (1980). Beyond conspecifics: Is Brer rabbit our brother? *BioScience, 30*, 763–768.

Cain, A.J. (1993). *Animal species and their evolution*. Princeton University Press.

Challenger, M. (2021). *How to be animal. A new history of what it means to be human*. Canongate Books Ltd.

"Crows are as smart as a seven-year-old child, researchers find" (2014, March 27). *Nature World News*. https://www.natureworldnews.com/articles/6450/20140327/crows-smart-seven-year-old-child-researchers-find.htm

Dolby, N. (2019). Nonhuman animals and the future of environmental education: Empathy and new possibilities. *The Journal of Environmental Education, 50*(4–6), 403–415. https://doi.org/10.1080/00958964.2019.1687411

Gardner, H. (1983). *Frames of mind: A theory of multiple intelligences*. Basic Books.

Herzog, H. (2010). *Some we love, some we hate, some we eat*. Harper Collins Publications.

International Commission of Zoological Nomenclature. (2012). iczn.orghttps://www.iczn.org/the-code/the-code-online/

King, B.J. (2021). *Putting compassion to work for animals in captivity and in the wild*. The University of Chicago Press.

Menzel, R. (2012). The honeybee as a model for understanding the basis of cognition. *Nature Reviews Neuroscience, 13*, 758–768. https://doi.org/10.1038/nrn3357

"Pigs are gentle creatures..." (2023). *Pigs | The Humane Society of the United States*. https://www.humanesociety.org/animals/pigs

Robson, D. (2021, April 7). Human-like intelligence in animals is far more common than we thought. *New Scientist*. https://www.newscientist.com/article/mg25033291-700-human-like-intelligence-in-animals-is-far-more-common-than-we-thought/

Safina, C. (2020). *Becoming wild. How animals learn to be animals*. Oneworld Publications.

Scott, E., Kallis, G., & Zografos, C. (2019). Why environmentalists eat meat. *PLoS ONE, 14*(7). https://journals.plos.org/plosone/article?id=10.1371/journal.pone.0219607

Serpell, J.A. (2004). Factors influencing human attitudes to animals and their welfare. *Animal Welfare, 13*, Supplement 1, 145–151.

Sternberg, R.J. & Grigorenko, E.L. (2004). Intelligence and culture: How culture shapes what intelligence means, and the implications for a science of well-being. *Philosophical Transactions of the Royal Society B: Biological Sciences, 359*(1449), 1427–1434. https://doi.org/10.1098/rstb.2004.1514

Wynne, C.D.L. & Udell, M.A.R. (2013). *Animal cognition: Evolution, behavior and cognition* (2nd ed.). Palgrave UK.

ZOOLOGICAL EMOTIONAL SCALE

All animals are equal, but some animals are more equal than others.

(Orwell, 1951, p. 114)

In 1945, an adult novella with the unusual title of *Animal Farm: A Fairy Story* was published by Martin Secker and Warburg Publishers. Written by George Orwell as a political satire it told the story of a group of farm animals who rebelled against their human farmer in order to create a society where all the animals were equal, free and happy. Seven commandments were painted in large letters on the barn wall, the seventh of which simply stated: "All animals are equal". As the reality of equality slowly proved somewhat different to the dream, the animals awoke one day to find a contradiction to this simple commandment had been mysteriously added—"but some are more equal than others". The animals' idyllic plans collapsed into an oppressive regime, ruled by the manipulative pigs. Why pigs were chosen to represent the (human-like) despotic rulers is beyond the scope of this book, but it is noteworthy that even in the 1940s, pigs were perceived to be intelligent creatures, albeit evil in this scenario. By contrasting the phrase "all animals are equal" with "some animals are more equal than others" in the above quote, the reader is introduced to an example of meaningless political discourse that remains relevant today when considering the different versions of equality imposed on animals in human societies.

The brief examination in the previous chapter of some animal classification systems, ranging from the inflexible phylogenetic hierarchies to the more flexible sociological scales, provides insight into how and why animal equality is so contested. Whether for biological characteristics, intelligence, utility or geographic location, some animals are perceived as more worthy of ethical consideration than others within a human society. Drawing from, and building on these previous examples, the Zoological Emotional Scale provides a framework within which to explore the moving pieces of how people perceive emotional capacity in animals, both between and within species, and the welfare implications of these perceptions. The framework can then become a tool to enact understanding, critical reflection and the possibility of change within the human–animal relationship.

Four domains of animal classification have the potential to influence and guide a person's perception of an animal's emotional capacity and therefore their responses to that animal or animal group. Positioning a species, a designated group within a species or an individual animal as friend, foe, functional or based on fallacy can have quite dramatic consequences for the animal. These categories are not discrete or static,

DOI: 10.1201/9781003298489-3

and members of the same species may fall into one or more domains, depending on the perception, attitudes and beliefs of the humans with whom their lives directly or indirectly intersect. Cats, with their 9,000-year association with humans, offer a glimpse into the complexity of these multiple perspectives.

By an accident of birth, a cat may be positioned as friend, capable of experiencing happiness and attachment when living as a beloved companion animal and family member. This cat is individualised and awarded personhood, although still remaining property of the human. Other cats may fall into the foe domain as either disposable disease-ridden colony cats living and breeding in city streets and abandoned buildings, or ruthless feral murderers of native wildlife and to be routinely culled by poisoning. Foe cats are not individualised and remain a generic enemy. Still other cats may serve as a food source in some cultures, moving them straight into the functional domain. Barn cats also move to the functional domain in their capacity as rodent killers. Barn cats are supported to kill, as opposed to the feral cats who are actively culled for killing. All these cats share the same neural substrates capable of supporting conscious experiences, as identified by the *Cambridge Declaration on Consciousness*, yet only those in the friend domain are provided a position of relative security within human society and the freedom to exercise their emotional capacity. When applying the Zoological Emotional Scale to cats, it becomes evident that their welfare considerations are shaped by a convergence of competing perspectives related to their perceived emotional capacity and influenced by social and cultural constructions or their relative value compared to native wildlife.

An animal categorised as friend may be awarded human-like status, and the human-like emotions that accompany this. Friends are the animals that humans take into their home and their heart, including the companion animals who achieve family-member status; the farmed animals or wild animals liberated from their functional meat, product or entertainment status to live out their lives in an animal sanctuary; and the wild animals never met (such as elephants), but still perceived as having emotional capacity based on learned information in books, documentaries and social media. Fundamental to the friend domain is the human's perception of the animal as an individual—*someone* rather than *something*.

An animal categorised as foe is perceived as a nuisance or danger to humans and to be ignored or destroyed. Any emotions attributed to this animal are negative, non-consequential or non-existent, as evidenced by the regular killing of the hardy and poisonous Cane toad, introduced to Australia to control cane beetles. Cane toad killing in Australian backyards became a cruel family sport in some areas, consolidating the amphibian as a foe without feelings or worth.

An animal categorised as functional performs some role for human or environmental benefit, sometimes to the detriment of the animal's wellbeing. Ascribing emotions to functional animals such as chickens, mass produced and over-fed to service fast food outlets, can prompt cognitive dissonance and a need to minimise or eliminate any unpleasant feelings of conflict. Consumers who eat the end product and meat workers who slaughter them on a daily basis must perceive these chickens as functional emotionless commodities. Animals such as earthworms are also perceived as emotionless, despite being crucial to the environment. Their bodies house a nervous system, respiratory system and sense organs, but their alien characteristics and lack of scientific evidence to the contrary may negate perceptions of emotional capacity.

An animal categorised as fallacy occupies a role that may be based on hearsay, rumour and stereotyping rather than fact or first-hand experience. Emotions attributed to these animals can be erroneous, serving to reinforce pre-conceived ideas and anthropocentric assumptions at the expense of the animal's or species' wellbeing. Ethologist Frans de Waal (2001) provided insight into moving beyond the fallacy domain when he commented that immersing oneself in the emotional lives of animals was the best way to understand and appreciate their emotional capacity. "Closeness to animals," he wrote, "creates the desire to understand them, and not just a little piece of them, but the whole animal" (p. 41).

How an animal ends up in one or more domains can be dependent on a mix of factors, including the visual appearance of the animal; first-hand experience of the animal, sometimes originating in childhood; learned experience of the animal including vicarious learning, formal studies and media generated emotional responses; and implicit or explicit cultural beliefs. The Zoological Emotional Scale therefore is constantly shifting to accommodate dominant, sometimes conflicting, perceptions that may vary among individuals, communities and cultures.

VISUAL APPEARANCE OF THE ANIMAL

With eight hairy legs, eight eyes and the ability to strike terror into humans hundreds of times their size, fear of spiders is deeply embedded in the human psyche. This fear, labelled arachnophobia, is reinforced by social learning and horror movies, and yet spiders are essential members of natural ecosystems on almost every continent of the world (Figure 2.1).

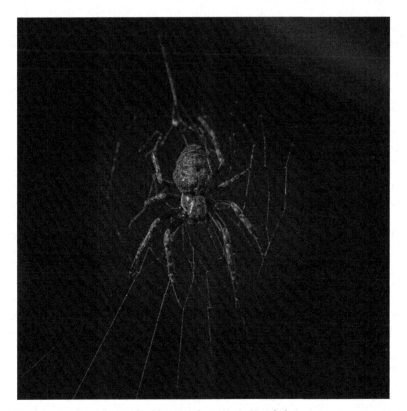

Figure 2.1 A spider, photographed by Florian Schmetz, via Unsplash.

Many people find it easier to relate to animals more like themselves, giving the warm-blooded mammals with two eyes a distinct advantage over spiders, insects and even fish and reptiles when it comes to attributing feelings of any kind. It is therefore unsurprising that the "us versus them" scenario is most evident in the lower tiers of the taxonomic hierarchy described in the previous chapter. As the lower-tier species diverge further from human-like characteristics, less human compassion is triggered compared to the relatable mammals higher in the hierarchy. This may extend to complete rejection of a species perceived as fearsome and disgusting, often with dire consequences for that species.

Biophobia is the human fear of certain species and a general aversion to nature that prompts a greater connection with technology and other human objects, interests and constructions rather than nature. Agrizoophobia describes a phobia for animals in general and can be further broken down into around 25 documented phobias related to particular animal groups. Insects (entomophobia or insectophobia), spiders (arachnophobia), bees (apiphobia or melissophobia), ants (myrmecophobia) and snakes (ophidiophobia) are the most common phobias (Castillo-Huitrón et al., 2020).

Human phobias have negative consequences for both humans and animals. Feelings of disgust reinforce phobias, leaving little space or opportunity to consider the animal as anything more than a catalyst for fear to be avoided or destroyed. While humans have the power to seek therapy or other strategies to manage their phobic feelings, the creatures themselves remain subject to the consequences of being feared and hated by a dominant species.

Anthropomorphism can also play a role in human perceptions based on visual appearance. Anthropomorphism is the attribution of human-like traits, emotions or intentions onto nonhuman entities, including animals, and is addressed in greater detail in Chapter 5. Humans find it easier to project feelings onto an animal with emotive eyes and human-like facial expressions or body language. A kitten has four hairy legs and two human-like eyes, making it a more worthy recipient of positive attributions compared to the Goliath Birdeater Tarantula spider of the same size. The iconic "Grumpy Cat" images that flooded social media demonstrate the ease with which humans can project feelings onto a face that meets similar specifications to human faces.

Context can change anthropomorphic projections, as evidenced by feelings about the feral cat population on a small island off the mainland of South Australia. The Kangaroo Island Feral Cat Eradication Program privileges the lives of native birds, small mammals and reptiles over the introduced feral cats decimating these island inhabitants, especially after the devastating 2020 bushfires (Hughes, 2021). Images of feral cats often include a small animal in their mouth, thus justifying traps that differentiate cats from native species and automatically spray toxins on the cats as they walk past. The cats are driven to groom the gel off their coats, thus ingesting a lethal dose. Perceived emotional capacity and positive anthropomorphic projections on these feral cats are over-ridden by the publicised pain, suffering and risk of extinction to native species. Feral cats still have the visual appearance of cats, but for many people, they are an abstract entity with which they have no first-hand experience, interaction or access, thus allaying some of the cognitive dissonance when animals within the same species are perceived and treated differently.

FIRST-HAND EXPERIENCE OF ANIMALS

To further confound consistency in the perceived emotional capacity of animals, first-hand experience and familiarity with an animal can influence a person's attribution of emotions. This starts early in life, with children less likely to assume moral superiority over animals and even to give more moral weight to animals rather than to humans (Herzog, 2021). Not surprisingly, studies suggest a person who grew up in a household with animals feels a greater emotional connection to living animals and perceives more similarities between human and animal emotions, a belief that could linger into adulthood (Menor-Campos et al., 2019). In fact, living with animals has been cited as the most important influence on reporting emotional capacity in animals (Morris et al., 2012). Conversely, never living with an animal can result in the attribution of far fewer emotions, irrespective of species (Knight et al., 2004).

Early in my studies of the human–animal relationship, I spoke with a woman who had left a rural property after her relationship broke down. That also meant leaving the horses she had loved and nurtured for many years. Head down and with lowered voice, she described overwhelming feelings of guilt and shame at deliberately choosing not to go back and visit these horses, compounded by the bitter accusations from her ex-partner of having never truly cared for these animals. As we spoke, it was evident that the reality was very different. She feared making the horses sad each time she re-appeared then left again, choosing instead to remove herself completely and allow the horses to move on without her. Having grown up with horses, the woman had no doubt about a horse's memory and positive and negative emotions, long before research caught up and confirmed the capacity for fear, pain, boredom, happiness and sadness in these sensitive animals.

First-hand experience can change a person's perception of emotional capacity in animals quite dramatically, especially when confronted by the harsh reality of factory-farmed animals. Mother cows calling for their newborn calves when forcibly separated to increase their milk yield for human consumption can be heart-breaking and leave no doubt that these mothers are mourning. But few people see this first-hand and should confronting undercover pictures be released to the public, the impact is often transient. Images may challenge the morals of those viewing and briefly trigger cognitive dissonance at this stark reminder of animal suffering for a glass of milk, but viewers also have the power to turn away from second-hand suffering. In contrast, philosopher, ethicist and author Lori Gruen (2014) described being "radically transformed" after befriending a young chimpanzee named Emma. Gruen's first-hand interaction with this single animal facilitated an awareness of the harms being done to chimpanzees and prompted Gruen's advocacy for chimpanzees all over the world.

Human attitudes towards arthropods (for example, insects, spiders, mites and ticks) present an anomaly as first-hand experience is not lacking in many households, gardens, produce farms and natural settings. This familiarity may be insufficient to circumvent a person's fear and aversion, blocking any consideration of the creature's feelings in the perceived need to destroy them. Many arthropods are thus firmly positioned as foe, despite their functional capacity within ecosystems. The visual appearance of arthropod bodies can also alienate humans. While the front segment of some insect bodies may be similar to mammals with mouth and eyes, similarities end there. Eyes can be numerous and scattered all over their body, thus creating a sense of "otherness". The male swallowtail butterfly, for example, has an eye on its penis for ease of mating, while the female has an eye on her rear end to check egg laying (Sverdrup-Thygeson, 2019). Swedish botanist Carolus Linnaeus believed insects did not have brains, a belief perpetuated by the fact that some insects could live for several days after being beheaded. This reinforced a sense of difference, even though Linnaeus's belief was later shown to be erroneous with insects having both a brain and a nerve cord running through their entire body with "mini brains" in each joint. An enduring and iconic Australian advertising jingle for bug killer spray created over 50 years ago provides an interesting contrast of familiarity, disgust and rejection coupled with the attribution of emotions to a common housefly. The catchy jingle about "Louie the fly" depicts him as a cheeky, endearing gangster-like individual who needs destroying due to his link with dirtiness and disease spreading, despite being the apple of his doting mother's eye.

As humans have become more disengaged from nature in some societies, knowledge of animals has increasingly been derived from formal and informal learning, vicarious learning and the media rather than first-hand experiences and familiarity. Learned attitudes to animals can be both positive and negative, but they also have the potential to be unlearned and re-shaped.

LEARNED ATTITUDES TO ANIMALS

In the 1970s, irrational fear of sharks escalated following the release of a book and movie simply titled *Jaws*. "Smashing together, they crush bones and flesh and organs to jelly. The jaws of a giant killer shark that terrorises a small holiday resort on Long Island…", states the descriptor in publisher Pan Books' 1975 edition of Peter Benchley's book. The movie, released in 1975, employed mechanical sharks in conjunction with an eerily ominous musical theme that induced terror by simply hinting at the shark's impending presence.

Sharks play an important role in their natural environment, and while many narratives following death or injury to humans still portray a fabricated image of implacable and voracious predators, a 2021 report on public attitudes to sharks suggested this may be changing with growing knowledge (Giovos et al., 2021). Among the formal and informal education modes, documentaries were a primary source of information about sharks, followed by the web, books and non-government organisations. Cumulatively, these went some way to buffering the perception of sharks as cold and cruel monsters of the sea created by books, movies and media reports. In fact, a study seeking public perceptions of charisma in animals positioned the great white shark as 14th on the list of the top 20 most charismatic animals (one above crocodiles and two above dolphins). Traits associated with charisma in the study included Rare, Endangered, Beautiful, Cute, Impressive and Dangerous, with the shark being described by all traits—except cuteness (Albert et al., 2018).

Public perceptions of sharks may be changing through education, but they can still linger based on first-hand experience. When speaking with an Australian surfer following his close encounter with a shark, he described the animal as powerful, beautiful and awe-inspiring, but with expressionless eyes and indecipherable facial expression. Instead of portraying the shark as large, he used the more derogatory descriptor of monster, all the while acknowledging that the shark had a stronger claim to exist in the sea than he had. When asked about perceived emotional capacity within the shark, the surfer hesitated, and then with a laugh replied, "There was definitely hunger for me…".

Perceptions and beliefs about the emotional abilities of elephants provide another example of acquiring attitudes through learning. Ranking third on the charismatic scale mentioned above, there is no shortage of documentaries and books on the emotional lives of elephants (see, for example, Carl Safina; J. Moussaieff Masson and S. McCarthy; and Cynthia Moss). As evidence of the emotional capacity of elephants accumulated through documentaries, personal narratives and books, I was privy to the confessions of a man whose learning about elephants had prompted feelings of guilt and regret. His past enjoyment of elephant rides on regular trips to Asia abruptly ended as he learned of the emotional lives of these family-oriented animals and the cruelty of forcing them to entertain tourists. According to World Animal Protection (2020), thousands of elephants are used and abused to give tourists a once-in-a-lifetime holiday experience, including the popular elephant washing. The physical and psychological suffering behind forcibly subduing these wild animals is immeasurable. As we chatted, the man expressed disgust at having once enjoyed this activity without thought for the feelings of the exploited animal upon whom he so thoughtlessly sat. Despite first-hand experience of elephants in an artificial and human-centric environment, his learning through secondary sources was dominant. As he commented, "Looking back I can clearly see the suffering in that shackled animal's eyes, but I was oblivious in the pursuit of a good selfie".

As human domination continues to encroach on natural environments, elephants and other wild animals can shift into the foe domain when competing against domestic livestock, crops and other human infrastructures. First-hand experience in these cases can have lethal consequences for the offending animals as they are destroyed with no thought of their feelings or needs.

CULTURAL FACTORS

Inseparable from the influences of visual appearance, first-hand experience and learned attitudes are the implicit and explicit cultural learnings that may underpin a person's perception of an animal's role and their belief in the animal's emotional capacity. Some animals have contributed to human cultural development and gained admiration and respect as spiritual, cultural or religious icons. Animism, the worldview that perceives humans, animals, plants and inanimate objects as potentially endowed with a spirit, permeates many non-Western cultures and blurs the boundaries between human and animal feelings (DeMello, 2012). Cultural factors therefore can be a contributing factor to the perception of emotional capacity in some animals—and conversely, to the disregard of emotion in other animals and their relegation to functional or foe status.

In Australia, kangaroos, described "as sentient animals with rights, and as animals highly significant to Indigenous people" (Boom et al., 2012, p. 34), nevertheless occupy a contested space in human perceptions. Despite being of cultural significance to Indigenous peoples, a high-profile iconic emblem on the Australian coat of arms and a fluffy replica in souvenir shops, kangaroos simultaneously represent a pest to the pastoral industry when competing with livestock for resources, and a source of high-protein meat for both human and animal consumption. Kangaroos were systematically slaughtered, their habitat destroyed and their numbers reduced in some areas of Australia. Defining kangaroos as foe, despite being a native animal, allows cruel treatment and culling when numbers expand, irrespective of the cultural significance and perceived emotional capacity of this unique marsupial seeking to exist on land that was once their own. The four kangaroos in Figure 2.2 graze on semi-rural land that is slowly being overtaken by human housing and other structures.

Tradition and trade can merge with potentially disastrous consequences for the welfare of individual animals, as highlighted by the Yulin Dog Festival, otherwise known as the annual Lychee and Dog Meat Festival held in parts of China during the summer solstice. Cultural conditioning has normalised the consumption of dogs and cats in some

Figure 2.2 Four kangaroos graze on short grass in a semi-rural area. Photograph supplied by the author.

Asian countries, but this is increasingly being challenged locally and globally as dogs and cats consolidate their position as friends capable of full emotional lives. When rallying support to ban this festival, animal welfare organisation People for the Ethical Treatment of Animals (PETA, 2023) painted a poignant heart-breaking picture of the dogs' emotional capacities with prominent social media campaigns and statements such as

> Killing, cooking, and eating dogs and cats is unthinkable to most of us, because they're our family members and our best friends. We can imagine the fear that they feel when they're caged with other animals and the agony that follows when they're bludgeoned or even skinned alive before being turned into pieces of meat.

Drawing attention to the dog or cat's feelings of fear immediately after reminding the reader of their own beloved best friend creates synergy in the reader's mind and an evocative prompt to action. Calling for a ban on the eating of cats and dogs is indicative that traditional cultural practices and attitudes towards animals can change when the discomfort of cognitive dissonance is triggered by a mismatch of cultural learnings and a person's reality. The decision to discard or maintain cultural beliefs can influence perceptions of an animal's emotional capacity and subsequent treatment.

STAGES OF CHANGE

The Zoological Emotional Scale is not a static framework, rather a starting point within which to develop a greater understanding of the perceived emotional capacity of the animals with whom we share this planet. It provides a tool to facilitate behavioural change where a person is seeking to understand and re-align their relationship with the natural world. However, putting together all the moving pieces of animal emotion is only the beginning if change is to occur. As Melanie Challenger wrote, "And so, if we no longer see into the lives of other animals, it's not because they don't have minds or we can't. It's because we don't want to" (2021, p. 119).

Human behavioural change can be conceptualised as occurring in six stages. Each stage represents specific groupings of attitudes, intentions and behaviours that must be completed across time so as to facilitate movement to the next stage of change (Prochaska & Norcross, 2014). How long a person remains at each stage varies, and people may cycle back to earlier stages or cease the process altogether. Understanding these stages can help the process of changing attitudes to animals and the environment.

Figure 2.3 provides a visual representation of the stages through which a person travels to achieve behaviour change. During the pre-contemplation stage, a person has no intention to change in the foreseeable future. However, a growing awareness that a problem exists leads to serious thinking about overcoming the problem, but still no commitment to change in the contemplation stage. This thinking leads to an intention to act with some behaviour changes but no effective action as the person moves into the preparation stage. The action stage is achieved when a person demonstrates overt behavioural changes and commitment. Work to prevent relapse and consolidate gains pushes a person into the maintenance stage. Once there is no longer any temptation to return to the problem behaviour, change is complete and stable and the termination stage has been successfully navigated.

It is symptomatic of the dominance of humans over animals and the natural world that a model of human change is even mentioned in a book about animal emotional capacity. Humans have long sought to change animals, with sometimes disastrous consequences for the individual animal, the species and the natural world. This disempowering of animals and the privileging of human interests suggest change must come

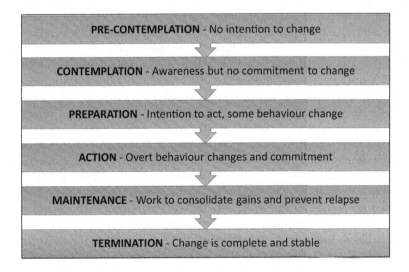

Figure 2.3 The Stages of Change—a model depicting each stage that a person can go through in order to establish a stable and lasting change in behaviour (Prochaska & Norcross, 2014).

from humanity in order to create a more equitable, sustainable and safe future. During the pre-contemplation, contemplation and preparation stages of change, a person increasingly seeks information to direct their musings and untangle feelings of cognitive dissonance. It is here that the Zoological Emotional Scale can provide a framework to represent the complexities and inconsistencies inherent in the way that humans perceive and treat animals.

Applying the Zoological Emotional Scale to rats allows a visual representation of this rodent's position in society and the subsequent impact on their wellbeing. This provides a personal tool to understand and interrogate the numerous perspectives of rats and their emotional capacity and dispute any cognitive dissonance arising. Spend some time interacting with a rat, and it is clear that they are individuals. They are curious, intelligent, have wonderful memories and show empathy to each other. Physiologically, rats share the same characteristics and needs, irrespective of their function and location, but their treatment by humans can be very different depending on their positioning within society.

Rats can be admired as survivalists, often thriving on the waste of human activity. These survivalist traits also see them eliciting revulsion through their perceived association with disease, squalor and death such as the Black Death pandemic of London. In New Zealand, introduced rats are considered one of the top three threats to native life, while in research facilities, rats are a valuable commodity where they are afforded death by humane killing, a benefit not extended to rats who infest human habitats or native wildlife areas. Rats are also a popular companion animal, described as loving, hygienic and capable of showing positive emotions when socialising or during playful handling and tickling. Tickling is increasingly being promoted as a means to habituate research rats to handling, thus reducing stress and minimising negative effects on experimental outcomes.

Visually representing this rodent on the Zoological Emotional Scale raises the question of how one small animal can occupy so many conflicting positions within a human-dominated world. Understanding how the human-assigned social role and perceived emotional capacity for any given rat, or group of rats, can determine their life and death experiences provides important information during the pre-contemplation, contemplation and preparation stages of change. When applying this framework, each person starts from their own point of familiarity with the animal and is prompted to consider "why do I think this way and are there other ways of perceiving this animal?" While this may create feelings of cognitive dissonance, it can also start the process of change to address negative stereotypes and outcomes for that animal.

Figure 2.4 represents some of the possible interactions that stereotypical rats of the genus *Rattus* (for example, Black rat *Rattus rattus* and Brown rat *Rattus norvegicus*) can have with human lives, and the rationale behind the rat's perceived emotional capacity based on these interactions. The weightings that a person applies to each domain will impact their perception of the rat as a friend with emotional capacity and therefore to be valued as a unique individual; a functional creature to be valued for what it provides humanity or the natural environment; a foe representing a threat with negative or no emotional capacity and to be destroyed; or a misrepresented creature based on fallacy. First-hand experience plays a role in the friend, foe and functional domains, although for very different reasons, and learned attitudes can influence the foe and fallacy domains.

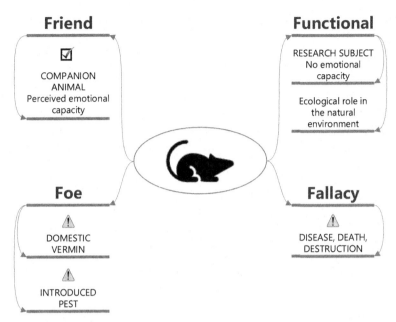

Figure 2.4 The Zoological Emotional Scale is used to identify where emotional capacity is attributed to the rat.

IMPLICATIONS

While it is not possible to consider every animal species in this book, the Zoological Emotional Scale is provided as a framework to give voice, albeit still a human voice, to the emotional world of animals. Each person will have differing and at times conflicting perceptions of animals, their rights and their place in a human-dominated society. Understanding how they arrived at those perceptions can move a person through the cycle of change and ultimately towards a less human-centric and more animal-friendly worldview.

Trying to categorise the natural world within any framework is reminiscent of Burghardt and Herzog's (1980) Play-Doh analogy where people such as animal activists who strive to change the status of animals in general or animal rescue workers who seek to save one animal at a time become over-stretched outliers to the dominant perception and treatment of some animal groups. Both activists and rescue workers can become overwhelmed as the sheer volume of animals in need threatens their physical and emotional resources whenever hard choices must be made. Befriending and attributing emotional capacity to animals in need, be they dogs, cats, farmed animals or wildlife, inevitably brings with it an emotional toll and feelings of helplessness and frustration about those animals left behind. As my father proved decades earlier, it is possible to shift individual animals from functional to the relative safety of the friend domain, but many more animals will never receive this opportunity. Without a major shift in the positioning of animals within a human-centric world, rescue operations will struggle to keep up with demand.

In 1974, philosopher Thomas Nagel posed a simple, but complex question: "what is it like to be a bat?" This question, and the quest to understand what animals think and feel, underpins cognitive ethology, a field originally criticised as being too subjective, anecdotal, anthropomorphic and therefore by default non-scientific. Decades passed and findings from rigorous naturalistic observation were increasingly replicated by laboratory-based researchers and sophisticated neuroscience techniques, prompting Marc Bekoff to comment in 2006 that cognitive ethology had become

> …the unifying science for understanding the subjective, emotional, empathic, and moral lives of animals, because it is essential to know what animals do, think, and feel as they go about their daily routines in the company of their friends and when they are alone.
>
> (p. 71)

The concept of cognitive ethology and the quest to understand what animals think and feel is embedded within the research, observations, interviews and case scenarios described in this book. Other key theoretical concepts addressed in the following chapters, including what it means to have consciousness, sentience and emotions, encourage the reader to further contemplate their professional or private relationship with animals and reconsider the troubled middle from multiple perspectives.

REFERENCES

Albert, C., Luque, G.M., & Courchamp, F. (2018). The twenty most charismatic species. *PLoS ONE, 13*(7), e0199149. https://doi.org/10.1371/journal.pone.0199149

Bekoff, M. (2006). Animal passions and beastly virtues: Cognitive ethology as the unifying science for understanding the subjective, emotional, empathic, and moral lives of animals. *Zygon, 41*(1), 71–104.

Benchley, P. (1975). *Jaws.* Pan Books Ltd.

Boom, K., Ben-Ami, D., Croft, D.B., Cushing, N., Ramp, D., & Boronyak, L. (2012). Pest and resource: A legal history of Australia's kangaroos. *Animal Studies Journal, 1*(1), 17–40.

Burghardt, G.M., & Herzog, H.A., Jr. (1980). Beyond conspecifics: Is Brer rabbit our brother? *Bio-Science, 30*, 763–768.

Castillo-Huitrón, N.M., Naranjo, E.J., Santos-Fita, D., & Estrada-Lugo, E. (2020). The importance of human emotions for wildlife conservation. *Frontiers in Psychology, 11*(1277), 1–11. https://doi.org/10.3389/fpsyg.2020.01277

Challenger, M. (2021). *How to be animal. A new history of what it means to be human.* Canongate Books Ltd.

de Waal, F. (2001). *The ape and the sushi master.* Allen Lane - Penguin Books.

DeMello, M. (2012). *Animals and society: An introduction to human-animal studies.* Columbia University Press. Giovos, I., Barash, A., Barone, M., Barría, C., Borme, D., Brigaudeau, C., Charitou, A., Brito, C., Currie, J., Dornhege, M., Endrizzi, L., Forsberg, K., Jung, A., Kleitou, P., MacDiarmid, A., Moutopoulos, K., Nakagun, S., Neves Flávia, J., Nunes, L.D., Schröder, D., … Mazzolid, C. (2021). Understanding the public attitude towards sharks for improving their conservation. *Marine Policy, 134.* https://doi.org/10.1016/j.marpol.2021.104811

Gruen, L. (2014). *Entangled empathy: An alternative ethic for our relationships with animals.* Lantern books.

Herzog, H. (2021). Why children would save animals over people. *Psychology Today, 54*(3), May/June, 48–51. https://research.ebscomedical.com/eds/detail?db=rzh&an=149936901

Hughes, M. (2021, December 30). Hundreds of feral cats removed from Kangaroo Island in bid to protect endangered native species. *ABC Rural.* https://www.abc.net.au/news/rural/2021-12-30/hundreds-of-feral-cats-removed-from-kangaroo-island/100730212

Knight, S., Vrij, A., Cherryman, J., & Nunkoosing, K. (2004). Attitudes towards animal use and belief in animal mind. *Anthrozoös, 17*(1), 43–62. https://doi.org/10.2752/089279304786991945

Menor-Campos, D.J., Hawkins, R., & Williams, J.M. (2019). Attitudes toward animals among Spanish primary school children, *Anthrozoös, 32*(6), 797–812. https://doi.org/10.1080/08927936.2019.1673055

Morris, P., Knight, S. & Lesley, S. (2012). Belief in animal mind: Does familiarity with animals influence beliefs about animal emotions? *Society & Animals, 20*, 211–224. https://doi.org/10.1163/15685306-12341234

Nagel, T. (1974). What is it like to be a bat? *Philosophical Review, 83*, 435–450.

Orwell, G. (1951). *Animal farm.* Penguin.

People for the Ethical Treatment of Animals (PETA). (2023). *Help animals like those suffering in the Yulin Dog-Eating Festival.* Action Centre - PETA Australia. https://www.peta.org.au/action/help-animals-like-those-suffering-in-the-yulin-dog-eating-festival/?gclid=Cj0KCQiAw9qOB-hC-ARIsAG-rdn6ed9gfXbIKzQbamVKhEh3mdRCS5UAHMEgg_HCHstK81xeboK7ybnQa-AsytEALw_wcB

Prochaska, J.O. & Norcross, J.C. (2014). *Systems of psychotherapy: A transtheoretical analysis* (8th ed.). Cengage Learning.

Sverdrup-Thygeson, A. (2019). *Extraordinary insects.* Mudlark.

World Animal Protection. (2020, August 12). *Tourists choosing elephant bathing over elephant riding, unaware of cruelty involved.* https://www.worldanimalprotection.org/news/tourists-choosing-elephant-bathing-over-elephant-riding-unaware-cruelty-involved

CONSCIOUSNESS, SENTIENCE AND EMOTION

The black dog crouches in absolute misery as a puddle of yellow urine snakes its way across the floor of the veterinary hospital. In stark contrast, a large smiling golden retriever sends out waves of happiness and floating hair with every sweep of her plumed tail. In the far corner, a Chihuahua shakes convulsively on his human's knee while simultaneously baring pointed teeth in a rictus snarl. An exuberant puppy, not yet clued in to the meaning of veterinary smells, stares gleefully at a caged cat, only to squeal in shocked disbelief when the cat swipes and hisses at the naïve youngster. On the other side of the room, a mournful yowl echoes eerily and attention is drawn to two glistening feline eyes of hatred in a darkened carry case. An embarrassed human is still trying unsuccessfully to alternatively coax and drag her terrified pony-sized dog through the door, while a parrot emitting intermittent screeches from a back room causes animals and humans alike to momentarily freeze in shock. Moving from the waiting room to the consultation rooms and hospital rooms, we find the large male cat who purrs so loudly during the examination that his heart rate and breathing are impossible to determine, while in the hospital recovery room two young kittens, inseparable since birth and recovering from their desexing operation, are inconsolable until placed together in the same holding pen. Spend a few hours in a busy veterinary hospital and it is impossible not to feel raw emotions pulsating through the room from the assortment of animals and humans.

Ten kilometres away in the waiting room of a veterinary clinic for "unusual" animals, it is a different story. The only tangible emotion in the room is human excitement at sharing snake and lizard anecdotes with like-minded people, tempered by some anxiety and fear over as-yet undiagnosed issues. Across the course of the morning, a small python, bearded dragon, tortoise, green tree frog and a fish with a large mass covering one eye silently wait in their individual carriers to be examined by the specialist veterinarians. The contrast between the palpable emotion visibly and loudly expressed by the warm-blooded patients and the seemingly emotionless and silent cold-blooded patients is stark.

These two waiting rooms replicate and solidify the perceptions that many people hold towards the emotional capacity of warm-blooded animals versus cold-blooded animals. Furry mammals and feathered birds offer the perceptible emotional responsiveness that many humans crave (de Waal, 2010), with dogs, cats and horses in particular

DOI: 10.1201/9781003298489-4

often described as being able to read human moods and reciprocate with empathy. Other vertebrate animals, including reptiles, amphibians and fish, as well as members of the Annelida Phylum (earthworms and leeches), Mollusca Phylum (snails and octopus) and Arthropoda Phylum (insects, spiders and crustaceans), are less likely to attract attributions of emotional capacity. In fact, visit a pet supplier that sells live crickets for reptile consumption and these insects are positioned as no different to buying a can of dog food or bag of cat biscuits. One notable difference is that purchasers are warned to only feed enough crickets that can be eaten immediately for fear that any excess crickets may escape and establish breeding colonies under feeding containers or in household plants. Despite being a reminder that these are living creatures, they remain commodified as food to be dusted with vitamin and mineral powder and offered to a reptile in the hope of simulating hunting instinct in the captive animal.

Contributing to the division between furry and feathered animals, and the scaly, slimy or creatures with exoskeletons is the emotive faces of mammals and birds, many of whom can move their eyes, mouths, ears, nose and beaks with purpose and meaning. Whether the lack of perceivable emotion makes it easier to commodify reptiles, fish and invertebrates is a complex question, supported by some horrific examples of poor treatment of these groups. In some countries, lobsters continue to be boiled alive and animal markets sell pieces of turtle meat freshly sliced from a living creature. Bees, already at risk from the insidious Varroa mite, face potential death when transported thousands of kilometres in trucks to provide year-round pollination of commercial crops. Of course, mammals are not immune to ill treatment despite their emotive faces. Rhinoceros horns have been cut from still living animals by poachers, just as shark's fins are sliced off living creatures and the shark thrown overboard to bleed to death or die of starvation at the bottom of the sea. Bears are trapped in tiny cages for decades to allow ease of access when draining their bile and research increasingly shows that companion animals may bear the brunt of human abuse as a perpetrator seeks to hurt another human through this vicarious violence.

The complexity of seeking to understand emotional capacity in animals is daunting but necessary work if humans are to cease misconstruing animals' lived experiences to the detriment of the individual, the species and the environment. Re-visiting the animal hospital waiting rooms mentioned earlier, it is important to clarify some important concepts when trying to make sense of the multitude of reactions and behaviours witnessed at these venues. In his seminal book *Beyond Words: What Animals Think and Feel*, Carl Safina (2015) offers a simple yet concise starting point to demonstrate the interconnectivity of some key concepts. According to Safina, consciousness is "the thing that feels like something" (p. 21); sentience is the ability to feel sensations, on a sliding scale from complex to seemingly none; thought allows consideration of something perceived; and emotion is the feelings around perceptions.

CONSCIOUSNESS

In 2012, the *Cambridge Declaration on Consciousness* formalised a scientific consensus that humans are not the only conscious beings. It stated that nonhuman animals—including all mammals and birds, and many other animals including octopuses—have neural substrates capable of supporting conscious experiences and awareness of their surroundings. Biologist Jonathan Birch concisely summed up consciousness with the statement that a conscious being has subjective experiences of the world and their own body (Birch et al., 2020).

What may seem common sense in the twenty-first century was not so well received in the 1970s when professor of zoology Donald Griffin created controversy with the contention that human mental experiences were not the only kinds that could exist. In his ground-breaking book, *The Question of Animal Awareness: Evolutionary Continuity of Mental Experience,* Griffin stated: "The belief that mental experiences are a unique attribute of a single species is not only unparsimonious; it is conceited" (1981, p. 170). Griffin was proved correct as the consensus achieved in the Declaration shifted the debate from whether nonhuman animals were conscious to what form their conscious experiences and awareness might take.

Drawing on neuroscience, evolutionary biology, comparative psychology, animal welfare science and philosophy, animal consciousness researchers set out to study the inner lives of animals and their subjective experiences in a scientifically rigorous way, modelled on the scientific study of human consciousness. With animals unable to verbally report their conscious experiences, these researchers searched for an integrated range of behavioural, cognitive and neuronal criteria for attributing conscious states. Cognitive ethology which combined two separate traditions—ethology and cognitive science—provided a missing component to this exploration through consideration of thought processes, beliefs, rationality, information processing and consciousness.

Jonathan Birch and colleagues (2020) adopted a conceptual multidimensional framework of consciousness that allowed the conscious states of animals to vary continuously along many different dimensions. They highlighted five significant areas of consideration: perceptual richness that varied in the detail with which animals perceived their environment; evaluative richness related to the ability to perceive an experience as good or bad; integration at time describing the perceived unity of a conscious experience; integration across time where experiences were perceived as a continuous flow; and self-consciousness that separated awareness of self from other aspects of the world. This provided the means to replace existing hierarchies that positioned some animals as more or less conscious than others with a multidimensional consciousness profile allowing for variation between species. Subjective life experiences and environmental factors were linked to degrees of consciousness across these dimensions, which in turn affected the complexity of any emotional response. Subjective experiences did not necessarily require high levels of brain organisation as neuro–ethological data suggested subjectivity could be linked to activity in the evolutionarily old parts of the brain.

The term consciousness has also been used descriptively by authors such as Melanie Challenger (2021) when citing examples of consciousness among whales who dream and wolves who carry mental maps in their heads. Bee expert Lars Chittka (2022), when examining the fascinating world of bees, described them as smart conscious beings, with distinct personalities and abilities to recognise flowers and faces, exhibit basic emotions and solve problems—in other words, possessing consciousness, albeit not modelled on human consciousness. Questions about the conscious experience of insects have long intrigued researchers. Bees come in a range of different sizes and all have much to lose if they damage their fragile wings when attempting to navigate a narrow space. Researchers Ravi and colleagues (2020) discovered that narrower gaps tended to elicit more concentrated "peering" by bees than wider gaps, and the larger bees with broader wingspans spent even longer peering. Grasshoppers and fruit flies also seem to know their own body and dimensions when navigating the environment, something that in humans is considered a component of self-awareness and linked to other emerging self-aware behaviours in children (Krause et al., 2019).

SENTIENCE

Imagine being dropped alive into a pot of boiling water. No doubt you would twist and squirm and scrape the sides of the pot in a desperate attempt to escape the pain during the long minutes it took to die—just like lobsters do. Conduct a quick search on the internet about how to cook lobsters and you will find many recipes still specify live lobsters and plunging them headfirst into salted boiling water. There are recommendations to choose lively lobsters, and warnings that lobster meat is not edible if it is not alive at the time of cooking. In 1973, a small article appeared in *The New York Times*, entitled "Easing the Lobster's Lot". Gordon Gunter, from the Gulf Coast Research Laboratory, suggested another method of killing these live animals without pain. This required the gradual heating of a pot of cool fresh water so as to anaesthetise the lobster through a combination of fresh water and slow temperature rise, as well as a small wire lattice on the bottom of the pot to avoid direct contact with the heated metal.

Despite Gordon Gunter's inference that lobsters could feel pain, the assumption that their simple nervous systems prevented this possibility lingered for decades and they were denied the status of sentience. Increasing evidence that this was a false belief resulted in some countries including Switzerland, New Zealand and Norway, legally banning the practice of throwing these live creatures into boiling water, while in 2022 the United Kingdom House of Commons passed an amendment to the *Animal Welfare (Sentience) Act* (2022). Crabs, lobsters and octopus were recognised as sentient beings in government policy, meaning restaurant patrons could no longer choose a live lobster from a prominently displayed tank to be boiled alive and served up with side dishes.

Biologist Jonathan Birch (2021) whose testimony contributed to this amendment provided a concise definition of sentience as an animal's capacity to have feelings with positive or negative quality. These included feelings of pain, pleasure, comfort, discomfort, boredom, excitement, contentment, frustration, anxiety and joy. John Webster (2007), veterinarian and Emeritus Professor of Animal Husbandry, was more succinct when commenting that sentience was feelings that matter.

Sentience builds on the subjective experiences of consciousness to include feelings about these experiences. No two sentient beings will have identical subjective experiences, but as a group they remain distinct from anything that is deemed to have no consciousness. Acknowledging and accepting vertebrate and invertebrate animal sentience requires moving beyond the perception that language is required to convey feelings, a sentiment articulated by Rene Descartes in 1646 and still hard for some to discard today. Descartes believed no nonhuman animal was sentient because the only evidence of sentience lay in the use of language to reveal an awareness of what things mean (Veit & Huebner, 2020). By setting this erroneous boundary, neither the bee who peers to work out if they can fit through a narrow gap nor the lobster who twists, turns and scrabbles when dropped alive into boiling water is sentient because they cannot articulate what they are thinking or feeling. While research has moved beyond the need for language, Birch warns that the urge to define feeling in terms of species-specific mechanisms should be approached carefully—or resisted—especially among invertebrates where there is an urgency to more fully understand sentience in the face of increasing commercial farming plans.

In 2021, a *Review of the Evidence of Sentience in Cephalod Molluscs and Decapod Crustaceans* proposed specific requirements for assessing sentience, based on the authors' empirical and theoretical expertise in animal behaviour, comparative cognition, sensory ecology, neuroscience, animal welfare and philosophy (Birch et al., 2021). In order to be deemed sentient, they believed an animal needed to possess nerve cell endings that facilitated the sensation of pain; integrative brain regions with connections between these regions; evidence of being affected by local anaesthetics or analgesics, including valuing analgesics when injured; motivational trade-offs between the cost of threat and the potential benefit of obtaining resources; flexible self-protective tactics in response to injury and threat and learning that went beyond mere habituation and sensitisation.

The ease with which humans relate to mammals and birds meant there was little dispute about perceiving these criteria and awarding sentience status to the familiar warm-blooded animals. However, some animals, described as the "awkward creatures", can be harder to engage with because they bite or sting, repulse or appear alien in terms of size, physiology, aesthetics and social organisation. These awkward creatures have elicited less scientific research that can influence perceptions of sentience with a resultant flow-on effect to their treatment, including the long-standing acceptance that it was appropriate to boil lobsters alive. The authors of the sentience review confirmed that when adopting their recommended criteria, they found very strong evidence of sentience among octopus, strong evidence in true crabs (belonging to a group called the Brachyura), and substantial evidence for squid, cuttlefish, anomuran crabs, astacid lobsters and crayfish and caridean shrimps. They pointed out that the amount of scientific research among a particular species will impact evidence of sentience, and where there is an absence of evidence through the absence of scientific research, this need not equate to the absence of sentience.

The experience of pain, one of the criteria for sentience, has been the focus of research among insects since the 1970s and 1980s. In 1979, Alumets found that earthworms possessed B-endorphins and enkephalins, naturally occurring peptides that modulate pain, resulting in the inference that they must experience pain to require this buffer. Insect physiologist Vincent Wigglesworth (1980), after extensive observation and reasoning, acknowledged that insects may not feel pain related to epidermal damage, but they did experience visceral internal pain as well as pain related to heat and electrical shock. Eisemann et al. (1984) recommended anaesthetising insects to avoid pain, while simultaneously concluding that there was insufficient evidence to support the experience of pain in insects in the same way that humans experience it. This all-pervasive anthropocentric tendency to use humans as the gold standard for experiences is fundamentally flawed when considering the many different pain experiences and tolerance levels evident among humans.

A systematic review of sentience among insects reported a greater focus on cognition (that is, acquiring knowledge) compared to sentience (Lambert et al., 2021), supporting the notion that both pain and feelings in these awkward creatures may be harder to acknowledge or accept. Stress, which was identified in two insect orders—Orthoptera (for example, grasshoppers, crickets and locusts) and Hymenoptera (for example, bees, wasps and ants)—can also be harder to disentangle from human standards.

Figure 3.1 A cricket missing a back leg was found in a domestic household space. Photograph supplied by the author.

Understanding stress in Orthoptera may prove essential as their numbers dwindle with potentially cataclysmic consequences for ecosystem functioning and vertebrate animal wellbeing (Millman, 2022). Crickets, like many invertebrates, have the capacity to regenerate parts of their bodies if damaged, perhaps contributing to the perception that insects do not feel pain. The cricket in Figure 3.1 is missing one of its large back legs that are crucial for jumping, thus placing the cricket at a greater risk of predators. The cricket was found in a domestic household space where it became a foe and faced an even greater risk from pest spray.

CASE STUDY IN COGNITIVE DISSONANCE

Drawing on Webster's contention that sentience comprises feelings that matter, the examination of two marine animals provides insight into why feelings may matter more in one creature compared to the other. Positioning the lucrative global tuna market and the marine mammal dolphin within the Zoological Emotional Scale sees a marked difference in perceptions of feelings in these two creatures that can have long-term welfare implications.

Tuna, rich in Omega-3, minerals, proteins and vitamin B12, has high global demand. During the 2020 COVID-19 pandemic lockdowns, consumption of canned tuna increased considerably, with tuna accounting for up to 20% of marine fishing and placing them at risk of being fished at biologically unsustainable levels (United Nations Environment Programme, 2021). Like all marine animals, tuna play an important role in the ecosystem and their over-fishing will necessarily have implications above and below them. Tuna fit somewhere between warm-blooded and cold-blooded creatures, with regional warm-blooded traits that allow them to warm parts of their bodies such as muscles, eyes and brain (Harding et al., 2021). They can also feel fear and pain, two evolutionary experiences that are evident in the frantic struggling of tuna as they are trapped in fish aggregation devices (FADs), pulled from the water and start to suffocate while being clubbed, stabbed with harpoons or slowly frozen. Researchers have known for years that fish are able to process noxious stimuli centrally, even without the neocortex found in mammals, and show a preference for areas of a fish tank with pain-relievers if hurting (Balcombe, 2016).

The use of FADs that encircle tuna with massive purse seine nets allows sufficient numbers of tuna to be caught to meet global demand. They also result in a bycatch of other marine species including sharks, rays and turtles as well as the dolphins under which tuna tend to congregate. To curb this accidental slaughter of dolphins, some nations such as United States enacted marine mammal protection acts to prohibit the killing, harassment and capture of marine mammals, further prompted by events such as the 1980s United States consumer boycott of tuna once dolphin deaths became public knowledge. "Dolphin safe" tuna labelling became common practice to placate consumer outrage over the marine mammal deaths (Allen, 2021). Privileging dolphins while ignoring the fear and pain experienced by the tuna prior to death reveals a human-imposed hierarchy of perceived sentience among these co-habitants of oceans, reinforced by media reports such as "Bottlenose dolphins are being caught and killed in trawl nets in Western Australia's north at unsustainable levels, a study warns" (Ramsey, 2022).

Dolphins live in the sea like tuna, but they are warm-blooded, air-breathing mammals who give birth to live young and nourish them with milk. They are also considered one of the most intelligent animals, demonstrating perception, communication and problem-solving. There are countless anecdotes of dolphins allegedly "saving" human swimmers, further endearing these marine mammals with their expressive faces, broad grin and loud vocalisations to the general public. Despite them also living in an alien watery environment, people seem to identify with dolphins as fellow-emotive creatures compared to the vitamin-rich tuna food source. One disadvantage dolphins face is the potential exploitation for human entertainment because of their intelligence, sociability, memory and personality, the very traits that privilege these animals over their non-mammalian marine co-habitants. Confining a social creature that would normally swim up to 60 km a day in a human-made pool to perform repetitive activities inevitably leads to negative feelings and behaviours.

Examining each of these sentient creatures within the Zoological Emotional Scale identifies a number of subjective distinctions to explain the preferential treatment of one animal over the other. Without these differing perceptions, cognitive dissonance would arise when selecting cans of tuna because of the dolphin-safe labelling while ignoring the suffering and painful deaths of the tuna within the cans.

Figure 3.2 represents some of the possible interactions that dolphins and tuna can have with human lives and provides some reasons for the different attributions of feelings to, and treatment of these two marine animals. The weightings that a person applies to each domain will impact their perception of sentience in these creatures and therefore their value as a friend and unique individual; or a functional creature to be valued for what it provides humanity or the natural environment; or a foe representing a threat with negative or no emotional capacity; or a misrepresented creature based on fallacy.

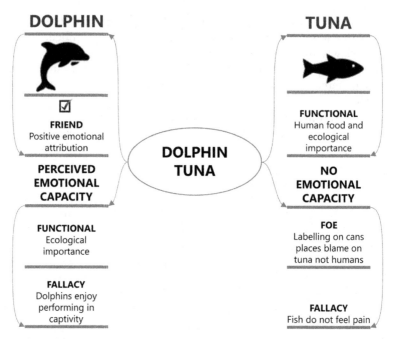

Figure 3.2 The Zoological Emotional Scale compares perceptions of emotional capacity in dolphins and tuna.

For dolphins, the weightings tend towards friend, based on their visual appearance, mammalian features and perceived intelligence. Dolphins are also perceived as functional within their community and ecosystem. Both domains prompt attributions of positive feelings and behaviours. Unfortunately, the fallacy that dolphins enjoy performing for human entertainment persists, reinforced by the dolphin's smiling visage and joyful antics.

For tuna, there is little motivation to attribute feelings, as to do so would invite cognitive dissonance when eating a can of tuna or tuna steak. Based on visual appearance and learned attitudes and beliefs, including the fallacy that fish do not feel pain, tuna are positioned as a valuable source of nutrients for human consumption. The tuna's functional role in their natural ecosystem is increasingly at risk with over-fishing to satisfy human demand. Despite growing evidence that fish can experience pain, there remains a lingering belief that capturing and killing them in mass nets does not inflict negative experiences. Labelling cans of tuna as ethically caught to save dolphins positions tuna as a risk to dolphin wellbeing and deflects blame from humans as the instigator of this risk.

DEFINING EMOTION

While sentience is an animal's capacity to experience and respond to both positive and negative sensations, emotion generates feelings about the sensations being experienced. Emotions can give rise to physiological, behavioural, cognitive and subjective responses.

When philosopher Martha Nussbaum (2006) provided a guideline for animal welfare, she purposefully included emotion as an entitlement that should not be impeded or disregarded. Item 5 of the entitlements states

5. *Emotions*. Animals have a wide range of emotions [...] they are entitled to lives in which it is open to them to have attachments to others, to love and care for others, and not to have those attachments warped by enforced isolation or the deliberate infliction of fear.

(p. 397)

Accepting that animals have an entitlement to live as emotional beings has flow-on effects for addressing their emotional interests and makes greater understanding and acceptance of animal emotion a priority. However, this requires consensus on how to define this controversial concept.

The words emotion, mood and affect are often used inconsistently and interchangeably in both the human and animal literature, thus contributing to and fuelling the debate over which, if any, animals experience emotion. When examining emotion among animals—including human animals—researchers Elizabeth Paul and Michael Mendl (2018) identified some common features that positioned emotion as a multi-component (subjective, physiological, behavioural and cognitive) response to a stimulus or event of importance to the individual. Emotion has valence (either pleasant or unpleasant) and varies in activation/arousal and duration/persistence from fleeting to days. The type of emotion will depend on the emotion-producing event, which can be external or internal, and the importance to the individual's goals and wellbeing. This positions emotion as an evolutionary survival mechanism for all animals, not just human animals.

Put simply, there is no reason to doubt that the pig, liberated from a breeding crate and enjoying a mud bath for the first time in her life in the freedom of an animal sanctuary, is experiencing a positive emotion as she grunts and wallows in appreciation. We will never understand her subjective appraisal of this new environment and the new friends sharing her experiences, but the physiological and behavioural indicators suggest that the mud and being with other pigs have made her a very happy pig (Figure 3.3). If we can see happiness in this pig, it raises the question as to why we cannot see, or choose not to see, misery in that same pig trapped in a breeding crate. The answer may be simple—to recognise misery is to invite guilt at knowing an animal is subjected to this, and to invite guilt is to create the mental anguish of cognitive dissonance.

Moods are linked to emotions and can arise from an aggregate of emotional experiences (Kremer et al., 2020). For example, fear and anger can leave animals or humans in a "bad mood", while positive experiences shift them back into a "good mood". Contributions of neuroscience and ethology to the study of emotion have broadened the evidence-base of animal emotion but have also created an important distinction between descriptive and prescriptive definitions of emotion and how that translates across animals and humans (Paul & Mendl, 2018).

A descriptive definition is like a dictionary, comprehensively setting out the ways in which a word is used in a given language and in everyday life. Descriptive definitions of emotion therefore risk becoming cultural constructs when subjectively applied to animals who may manifest emotion in their own unique, nonhuman way.

A prescriptive definition prescribes what elements are included when developing a scientific theory of a concept. In the case of emotion, this includes the basic biological structures and processes not acknowledged in a descriptive definition (Paul & Mendl, 2018). Neuroscience has clearly demonstrated that the anatomical and neurochemical systems underpinning core emotions are shared across many species—especially

Figure 3.3   Two pigs stand nose to nose in a field. Physiological and behavioural indicators show that being with other pigs make them happy. Photograph taken by the author at Edgar's Mission Farm Sanctuary, Lancefield and reproduced with their permission.

mammal species—making it clear that some animals have the capacity to experience emotions (Turnbull & Barr, 2020). Edmund Rolls' (2014) prescriptive definition views emotions as those states elicited by rewards and punishments and therefore part of instrumental learning. A reward is to be worked towards, and a punishment is to be escaped or avoided. This conceptualisation of rewards and punishments allows for measurable outcomes as well as different emotional responses based on individual neural systems specialised to the environmental and ecological niche in which the animal exists. Emotions and the subjective experiences prompting these emotions are not necessarily homologous with human emotions as different species evolved different approaches to dealing with the positive and negative stimuli within their environment.

Researchers David Anderson and Ralph Adolphs (2014) suggested four building blocks for describing different approaches to dealing with positive and negative stimuli: valence (whether the stimulus is good or bad); scalability (a graded nature of the response); persistence following stimulus cessation; and generalisation to different contexts, all of which are evident in human emotions. Animals who show evidence of these four building blocks can be labelled as having potential for emotion in a broad sense, allowing emotional possibilities to permeate many of the lower positioned phyla, including reptiles, fish and insects, and reinforcing the welfare issues when emotion is discounted and ignored in these groups.

AFFECTIVE NEUROSCIENCE

The term affect is often used interchangeably with emotion or mood in the human and nonhuman animal literature. Neuroscientist and Psychobiologist Jaak Panksepp (2004) introduced the term affective neuroscience in the 1990s and this became an accepted research area in cross–species brain science. Panksepp identified seven primary emotional systems called Seeking, Care, Play and Lust (positive affects) and Fear, Sadness and Anger (negative affects) which he believed evolution had endowed mammalian brains with as survival guides. Panksepp described each of these emotional systems as evaluative, providing a valence that is either pleasant or aversive and signalling which objects or situations to approach (seeking, lust, care and play) and which to avoid (rage, fear and panic).

Examined within the context of an animal's daily life, these primary emotions motivate an animal to maintain social contact and social bonding to avoid separation anxiety (Panic); identify potential mates and reproduce (Lust); nurture any resultant offspring (Care); and engage in physical activities that further bond and provide training for adulthood (Play). The animal must also be motivated to forage for resources (Seeking); compete and defend those resources (Rage); and escape from or avoid harm (Fear).

Each one of these primary emotional systems has subcortical brain areas that activate behaviour. Panksepp (2010) commented that while emotions are initially stimulated by a limited set of unconditional stimuli, the arousal must outlast the triggering circumstances, and the emotions felt by animals would be "more raw" compared to the comparably more controlled emotions in some humans. Panksepp also identified secondary and tertiary emotions, defined as emotion resulting from memory and learning processes and emotion requiring higher cognitive abilities and the activation of the neo-cortical structures in order to act with intention. Shame following the primary emotion Rage has been described as a secondary-process emotion. Intrapsychic ruminations and contemplation requiring neo-cortical structures are tertiary-process emotions and by definition difficult to explore in animals other than humans. However, just because something is difficult to explore does not preclude its existence.

Modern affective neuroscience has revealed some specific brain structures involved in emotions, including core systems of the peri-aqueductal gray, hypothalamus, amygdala and the anterior cingulate gyrus and nucleus accumbens (Turnbull & Bar, 2020). As always, researchers cannot resist comparing humans and animals to provide a recognisable baseline, and in this case ongoing and ever-evolving comparisons of brain structure do indicate similarities in neural circuits and parts of the brain associated with affective experience and processing, albeit with variations depending on the neural structure of the phyla and species. For example, in humans and other mammals the amygdala, part of the brain's ancient limbic system, helps drive emotional reactions, memory and decision-making. In a fish's brain, the medial pallium seems to play a similar role (Balcombe, 2016). Disabling this region experimentally in fish results in changes in aggression like that seen in mammals subjected to similar treatment of the amygdala.

Neuroendocrine markers as potential indicators of positive and negative affect have also garnered increasing attention, with evidence that primary affect wired into the subcortical structures are anatomically and neurochemically similar in function across all mammals, and in many cases, across all vertebrates (Balcombe, 2016). Cortisol, colloquially known as the stress hormone that acts to regulate stress, has been shown to perform similar functions in mammals and fish, as evidenced by studies among genetically modified zebrafish with cortisol deficits. Positron emission tomography scanning and functional magnetic resonance imaging have also been used to further compare brain activity in humans and animals and map affective systems, with separation distress in some animal brains appearing very similar to human sadness systems (Kremer et al., 2020).

Two important observations arise from Panksepp and other researchers' work in the area of animal emotion. Firstly, primary emotions as identified by Panksepp require no learning. The subjective, feeling component of emotion arises from ancient subcortical structures which can learn and adapt to novel environmental experiences and produce complex and varied behavioural responses (Montag & Davis, 2020). Emotions thus manifest in different ways and levels of complexity depending on the animal's environmental needs, many of which may be incomprehensible to humans. Secondly, by operating at a subcortical level, there is considerable evidence that the primary-process emotional brain systems do not require the neocortex, removing emotional capacity from perceived intelligence. In humans, strong emotion can restrict cortical activation, suggesting that the neocortex is more instrumental in restricting and regulating emotion rather than adding qualitatively to the experience.

BEYOND NEUROSCIENCE

Introducing neuroscience into the examination of emotions in animals offers some scientific validity which in turn increases its perceived credibility. However, Marc Bekoff (2000) commented that it does not negate the need for a mix of this "hard" research with "soft" (anecdotal) research to gain a greater understanding of the sometimes inexplicable and more nuanced behaviour of animals. This requires looking beyond the observable physiological and behavioural changes arising from specific circumstances (for example, the freeze/fight/flight response associated with Fear and stimulation of the amygdala; or Anger associated with stimulation of the hypothalamus) to the subjective conscious experience arising from these subcortical structures (Dawkins, 2000). It also requires resisting the tendency to use theoretical knowledge of human emotional experiences to infer emotional states in animals, nor setting human emotion as the standard, thereby potentially ignoring the possibility that animals have unique capacities arising from different experiences that humans do not have. The logic of this is undeniable when considering many animals inhabit niches unlike those of humans, and thus face evolutionary challenges that humans will never experience or understand (Mikhalevich & Powell, 2020). However, in some situations, there can be benefits to adopting anthropomorphic perspectives as a starting point, a consideration that is discussed in more depth in Chapter 5.

The unique experiences of animals become evident when exploring sensation and emotion in squids and octopus, animals that evolved complex brains independently of vertebrates (Crook, 2021), but remained so visually different to humans and mammals as to elicit less consideration in the research field of emotion. Despite these physical differences, when viewers were invited into the undersea world of a tiny wild octopus in the 2020 documentary entitled *My Octopus Teacher*, many were left in no doubt of the brave little cephalopod's curiosity in exploration, excitement and apparent affection for film-maker Craig Foster. Anthropomorphism combined with the words of the narrator provided a lens through which to overcome the alien nature of this creature.

My Octopus Teacher provided one anthropomorphic mechanism through which to make assumptions about emotions in an invertebrate animal, but there remains a question mark as to what emotions might look like, and how they could be measured, in other invertebrates, reptiles and amphibians. In particular, there is still more work to be done in both the scientific and ethological studies of emotion, sentience and consciousness among insects as mass edible insect farming grows in popularity. With evidence that some insects have evolved emotion-like states mediated by the

Figure 3.4 A small turtle gazes at the camera with an inscrutable expression. Photograph supplied by the author.

same neurotransmitters linked to human emotions (Mikhalevich & Powell, 2020), this knowledge can have far-reaching moral and welfare implications.

Among reptiles and amphibians, evolutionary biologist Gordon Burghardt (2013) was adamant that behaviour and emotions could not be viewed and interpreted through a human-centric lens, especially as cold-blooded (poikilothermic) vertebrates function at different time scales, lack facial and vocal signals of internal states easily interpretable by humans, and often rely on sensory cues that cannot be evaluated against a mammalian framework. For example, the small turtle in Figure 3.4 gazes directly at the camera, but its inscrutable expression gives no indication of its internal state. Despite this, Burghardt cited examples of some reptiles, including monitor lizards, iguanas, tortoises and crocodiles, who seemed to recognise different people and even approach to make contact or be stroked. Importantly, they seemed to enjoy sensations such as flowing water and engaged in object play, an observation linked to evidence that emotions can be mediated by the limbic system that both mammals and reptiles share.

PLAY

While few people may have had the privilege of watching reptiles at play, young mammals provide a more common and entrancing sight as they frolic in the serious task of mimicking future life-saving behaviours, or just run and bounce for the sheer joy of living. Burghardt (2005) defined play as "repeated, incompletely functional behaviour differing from more serious versions structurally, contextually, or ontogenetically, and initiated voluntarily when the animal is in a relaxed or low stress setting" (p. 82). When applying this definition to reptiles, he described Nile softshell turtles batting around basketballs and plastic bottles, swimming through or moving around hoops, and playing tug-a-war with their carers and a hose; or monitor lizards who manipulate, shake and carry around objects including rings, plastic discs and buckets, and who seem to develop a close attachment to their carers by approaching them with rubbing and other tactile movements. Play fighting has also been identified among adult dart poison frogs who demonstrate harmless fighting with each other (Burghardt, 2013).

Over the years, I have listened to many people proudly describe their companion animals at play, accompanied by an abundance of images on their phones. One young man became so animated as he described his cat's antics that his long-term clinical depression temporarily lifted and his face came alive. Chuckling at the memory, he recalled how his cat would ambush him on the stairs by hiding behind a pillar, then jumping out and meowing as he walked past. The young man likened it to someone saying "Boo" and was sure that all the other residents similarly loved his cat's amusing antics.

Apart from someone with severe allergies, most people would respond positively to a playful cat on a flight of stairs, whether they had met the cat previously or not. Human relationships with domestic cats are bounded by familiarity and expectations established over thousands of years. It is possible they may not respond so positively to a rat jumping out to say boo. As identified in Chapter 2, relationships with rats and perceptions of their role in society took a different turn to domestic cats. And yet rats can be just as playful as cats. Neurobiologist Jaak Panksepp identified a neurochemical basis of playfulness in rats, with an increase in opioid activity in the rat's brain associated with playing and the resultant pleasure. A playful rat is no different to a playful cat, dog or human, and yet human perceptions can prompt a very different outcome, including exterminators and poison baits, to a rat on the stairs.

Human perception of animals and their behaviour is crucial to animal welfare as any misunderstanding or aversion can endanger individual animals and species. However, aversions to the awkward animals, such as spiders and snakes, may have an evolutionary basis, according to researchers at the Max Planck Institute for Human Cognitive and Brain Sciences in Germany (Hoehl, et al., 2017). Babies as young as six months old demonstrated physiological stress responses to images of snakes and spiders compared to flowers and fish, suggesting that even modern-day humans are hard-wired to feel fear and disgust at these creatures that may have once posed a threat to long-ago ancestors. Moving beyond this innate aversion to perceive sentience, emotional capacity and playful traits may prove impossible for some people, thus skewing their personal Zoological Emotional Scale when assessing some animals.

IMPLICATIONS

Academic definitions of the concepts of consciousness, sentience and emotion provide some direction in the quest to understand perceived emotional capacity in animals. However, they fall short of explaining the multifaceted decision-making processes employed by humans to determine the treatment of specific animals within and between species. An animal's inability to articulate their subjective experiences of the world and their own body (consciousness); their feelings of pain, pleasure, distress or harm (sentience); or their subjective feelings about their experiences and feelings (emotions), remains a stumbling block for some people. Instead, emotional attribution is often dependent on anthropocentric interpretations that privilege human needs over animal needs. To do otherwise can invite cognitive dissonance and a need to reduce the resultant painful internal conflict.

Revisiting the dogs and cats in the veterinary hospital waiting room described earlier in this chapter, several of Panksepp's primary emotions can be interpreted through a human lens of emotion simply by listening to and observing the animal's behaviour. The black dog in a puddle of urine shows fear, while the smiling Golden Retriever is both seeking companionship and wanting to play. The trembling Chihuahua covers fear with rage, while the exuberant puppy flips from play to fear. The two glistening feline eyes are aligned with rage in contrast to the fearful pony-sized dog being dragged through the doorway. The two young kittens in for desexing and inconsolable until housed together represent a seeking behaviour that was immediately recognisable and responded to within human parameters.

Step into the treatment rooms and another scenario plays out where the animal's emotional experiences and needs are juxtaposed against the rationalisation of ethical decision-making. Welfare science promotes that animals have feelings and that negative feelings should be minimised, and yet often the animal's feelings are inadvertently over-ridden as they become objects of medical intervention and as such, removed from the decision-making process (Donald, 2019). Bioethicist and writer Jessica Pierce (2019) draws attention to the need for veterinary medicine to engage more actively in the body of research around animal emotion and the subjective experiences of animals, especially in relation to interpretation and management of pain, assessment of quality of life and the autonomy of the animal patient. Recognition that physical pain can be compounded by the emotional pain of loneliness and stress is commonplace in human medical practice and can similarly have an impact in veterinary medicine. Loneliness can increase inflammatory responses and sensitivity to pain, raising concerns about the many domestic animals that are isolated and denied adequate social interaction.

Misinterpreting or denying animal emotion based on anthropocentric bias has welfare implications beyond the involvement of animals in their veterinary treatment. Visit a rescue shelter and there will be at least one dog or cat who cannot adjust to shelter life. They will withdraw physically and emotionally, urinate and defecate uncontrollably and resort to fearful aggression, thus rendering the animal unlikely to be rehomed and at risk of euthanasia. This can be exacerbated by societal stereotypes of certain dog breeds as vicious or cats in general as aloof. Social media and cartoon images often depict cats as complacent, selfish and manipulative, albeit in a humorous manner. While cats may not exhibit the same easily recognised emotional responses compared to dogs, their more exuberant domesticated partner, overlooking a cat's emotional capacity may contribute to human perceptions that they can be abandoned without negative emotional consequences. Acknowledging that animals experience emotions raises questions about their capacity for family bonds and the implications when these bonds are ruptured. These questions are explored in the following two chapters.

REFERENCES

Allen, L. (2021, April 28). The origin of the "Dolphin-Safe" tuna label. *Forbes.*

Alumets, J., Hakànson, R., Sundler, F. & Thorell, J. (1979). Neuronal localisation of immunoreactive enkephalin and B-endorphin in the earthworm. *Nature, 279,* 805–806. https://doi.org/10.1038/279805a0

Anderson, D.J. & Adolphs, R. 2014. A framework for studying emotions across species. *Cell, 157*(1), 187–200. https://doi.org/10.1016/j.cell.2014.03.003

Balcombe, J. (2016). *What fish knows: The inner lives of our underwater cousins.* Scientific America/ Farrar, Strauss and Giroux.

Bekoff, M. (2000). Animal emotions: Exploring passionate natures. *Bioscience, 50*(10), 861–870.

Birch, J., Schnell, A.K. & Clayton, N.S. (2020). Dimensions of animal consciousness. *Trends in Cognitive Sciences, 24*(10), 789–801. https://doi.org/10.1016/j.tics.2020.07.007

Birch, J. (2021). *Invertebrate sentience* [Video]. https://www.lse.ac.uk/news/news-assets/pdfs/2021/ sentience-in-cephalopod-molluscs-and-decapod-crustaceans-final-report-november-2021.pdf

Birch, J., Burn, C., Schnell, A., Browning, H. & Crump, A. (2021). *Review of the evidence of sentience in Cephalopod Molluscs and Decapod Crustaceans.* London School of Economics and Social Science. https://www.lse.ac.uk/news/news-assets/pdfs/2021/sentience-in-cephalopod-molluscs-and-decapod-crustaceans-final-report-november-2021.pdf

Burghardt, G.M. (2005). *The genesis of animal play: Testing the limits.* MIT Press.

Burghardt, G.M. (2013). Environmental enrichment and cognitive complexity in reptiles and amphibians: Concepts, review, and implications for captive populations. *Applied Animal Behaviour Science, 147*(3–4), 286–298. https://doi.org/10.1016/j.applanim.2013.04.013

Cambridge Declaration on Consciousness (archive). (2012, July 7). Written by Philip Low and edited by Jaak Panksepp, Diana Reiss, David Edelman, Bruno Van Swinderen, Philip Low and Christof Koch. University of Cambridge.

Challenger, M. (2021). *How to be animal. A new history of what it means to be human.* Canongate Books Ltd.

Chittka, L. (2022). *The mind of a bee.* Princeton University Press.

Crook, R. (2021). *Crook Laboratory.* https://crooklab.org/

Dawkins, M.S. (2000). Animal minds and animal emotions. *American Zoologist, 40*(6), 883–888. https://doi.org/10.1093/icb/40.6.883

de Waal, F. (2010). *The age of empathy: Nature's lessons for a kinder society.* Three Rivers Press.

Donald, M.M. (2019). When care is defined by science: Exploring veterinary medicine through a more-than-human geography of empathy. *Area, 51,* 470–478. https://doi.org/10.1111/ area.12485

"Easing the lobster's lot." (1973, January 26). *The New York Times,* 27. https://www.nytimes. com/1973/01/26/archives/easing-the-lobsters-lot.html

Eisemann, C.H., Jorgensen, W.K., Merritt, D.J., Rice, M.J., Cribb, B.W., Webb, P.D. & Zalucki, M.P. (1984). Do insects feel pain? - A biological review. *Experientia, 40,* 164–167.

Griffin, D. (1981). *The question of animal awareness: Evolutionary continuity of mental experience* (Revised and enlarged edition). William Kaufmann, Inc.

Harding, L., Jackson, A., Barnett, A., Donohue, I., Halsey, L., Huveneers, C., Meyer, C., Papastamatiou, Y., Semmens, J.M., Spencer, E., Watanabe, Y. & Payne, N. (2021). Endothermy makes fishes faster but does not expand their thermal niche. *Functional Ecology, 35,* 1951–1959. https:// doi.org/10.1111/1365-2435.13869

Hoehl, S., Hellmer, K., Johansson, M. & Gredebäck, G. (2017). Itsy bitsy spider…: Infants react with increased arousal to spider and snakes. *Frontiers in Psychology, 8*(1710). https://doi.org/10.3389/ fpsyg.2017.01710

Krause, T., Spindler, L., Poeck, B. & Strauss, R. (2019). Drosophila acquires a long-lasting body-size memory from visual feedback. *Current Biology, 29,* 1833–1841. https://doi.org/10.1016/j. cub.2019.04.037

Kremer, L., Klein Holkenborg, S.E.J., Reimert, I., Bolhuis, J.E. & Webb, L.E. (2020). The nuts and bolts of animal emotion. *Neuroscience and Biobehavioral Reviews, 113,* 273–286. https://doi. org/10.1016/j.neubiorev.2020.01.028

Lambert, H., Elwin, A. & D'Cruze, N. (2021). Wouldn't hurt a fly? A review of insect cognition and sentience in relation to their use as food and feed. *Applied Animal Behaviour Science, 243* (October). https://doi.org/10.1016/j.applanim.2021.105432

Mikhalevich, I. & Powell, R. (2020). Minds without spines: Evolutionarily inclusive animal ethics. *Animal Sentience, 29*(1). https://doi.org/10.51291/2377-7478.1527

Millman, O. (2022). *The insect crisis: The fall of the tiny empires that run the world*. W.W. Norton & Company, Inc.

Montag, C. & Davis, L.K. (2020). *Animal emotions: How they drive human behavior*. Brainstorm Books.

My Octopus Teacher. 2020. Directed by Pippa Ehrlich and James Reed. Netflix. IMDB. https://seachangeproject.com/my-octopus-teacher/

Nussbaum, M.C. (2006). *Frontiers of justice: Disability, nationality, species membership*. Harvard University Press. https://doi.org/10.4159/9780674041578

Panksepp, J. (2004). *Affective neuroscience: The foundations of human and animal emotions*. Oxford University Press.

Panksepp, J. (2010). Affective neuroscience of the emotional brain mind: Evolutionary perspective and implications for understanding depression. *Dialogues in Clinical Neuroscience, 12*(4), 533–545. https://doi.org/10.31887/DCNS.2010.12.4/jpanksepp

Paul, E. S. & Mendl, M. T. (2018). Animal emotion: Descriptive and prescriptive definitions and their implications for a comparative perspective. *Applied Animal Behaviour Science, 205*, 202–209. https://doi.org/10.1016/j.applanim.2018.01.008

Pierce, J. (2019). The animal as patient. Ethology and end-of life care. *Veterinary Clinics of North America: Small Animal Practice, 49*, 417–429. https://doi.org/10.1016/j.cvsm.2019.01.009

Ramsey, M. (2022, October 27). WA dolphin bycatch 'unsustainable': Study. *7news*. https://7news.com.au/news/wa-dolphin-bycatch-unsustainable-study-c-6620001

Ravi, S., Siesenop, T., Bertrand, O., Li, L., Dousot, C., Warren, W.H. Combes, S.A. & Egelhaaf, M. (2020). Bumblebees perceive the spatial layout of their environment in relations to their body size and form to minimize inflight collisions. *Proceedings of the National Academy of Sciences of the United States of America, 117*(49), 31494–31499. https://doi.org/10.1073/pnas.2016872117

Rolls, E.T. (2014). *Emotion and decision-making explained*. Oxford University Press.

Safina, C. (2015). *Beyond words: What animals think and feel*. Henry Holt and Company, LLC.

Turnbull, O.H. & Bar, A. (2020). Animal minds: The case for emotion, based on neuroscience. *Neuropsychoanalysis, 22*(1–2), 109–128. https://doi.org/10.1080/15294145.2020.1848611

United Nations Environment Programme. (2021). *Sustainable tuna fishing key to protect the species*. https://www.unep.org/news-and-stories/story/sustainable-tuna-fishing-key-protect-species

Veit, W. & Huebner, B. (2020) Drawing the boundaries of animal sentience. *Animal Sentience 29*(13). https://doi.org/10.51291/2377-7478.1595

Webster, J. (2007). *New trends in animal welfare*. ISAH: Tartu June 2007. https://www.isah-soc.org/userfiles/downloads/proceedings/Plenary_2007/JohnWEBSTER.pdf

Wigglesworth, V.B. (1980). Do insects feel pain? *Antenna, 4*, 8–9.

FAMILY LOVE OR INSTINCT?

Contemporary definitions of family no longer limit membership to biological relationships and the sharing of genes. Families of choice can be subjective and based on mutual emotional bonds, common values, goals and responsibilities for the wellbeing of each other. Families facilitate social learning through the dissemination of information stored in the brains of other members, thus spreading changes in behaviours and customs without needing to wait for genes to catch up (Safina, 2020).

Family life is not unique to human animals. Pair bonding, parenting, attachment and kinship are evident across a range of species as they demonstrate emotional capacity in their own appropriate way. Implicit in family life, whether they are families by blood or families by choice, is the ability to cooperate, reciprocate and understand roles, relationships and at times obtuse rules and regulations. This raises the perplexing question of whether animals can feel love and attachment as defined and socially constructed by humans, or whether they are responding to instinctual needs, motivations, neurotransmitters and hormones. The question seems almost irrelevant, given that humans are animals and subject to the same catalysts including the bonding hormone oxytocin and feel-good endorphins. As Melanie Challenger (2021) commented, humans giving each other love and support is not a condition of human exceptionalism, rather part of the same compulsion experienced by any other animal.

In human families, there is a social expectation that children are cared for up to a certain age, after which they strike out alone. Two-year-old children, helpless and at risk of separation anxiety, elicit feelings of nurturance and attachment in their carers to consolidate this social expectation and survival necessity. In the animal world, many two-year-olds are independent, even parents themselves, although there are some exceptions. Marine mammal killer whales (orcas) are one of those exceptions and when Tilikum, a young killer whale, was taken from his family at two years old, the consequences were heart-breaking.

TILIKUM

In 2017 Tilikum, the killer whale who inspired the 2013 documentary *Blackfish*, died at the young age of around 35 years. In the wild, killer whales live in complex and highly social family pods ranging from 20 to 50 members. The pods are organised around the females, including a matriarch who holds a wealth of experience and

DOI: 10.1201/9781003298489-5

knowledge accumulated over a lifetime that can span 80 years. Males stay with their mothers their entire lives, some even dying not long after their mother's death. However, this was not to be Tilikum's experience. According to the late marine biologist Ken Balcomb, founder of the Center for Whale Research (2023) in Washington State, US, separation from family can be traumatic. A killer whale's family is for life, providing security and identity.

Captured and removed from his family at the age of two years and forced to perform seven days a week at marine mammal theme parks, Tilikum experienced bleeding stomach ulcers, boredom and involvement in the death of three humans. He was a male child who should have remained with his mother his whole life, rather than being kept in prolonged captivity with an extreme level of sensory deprivation. Documented captivity effects on both humans and animals can include long-term activation of the hypothalamic–pituitary–adrenal axis, which is instrumental in mediating the stress response, repetitive and abnormal behaviours called stereotypies and indicative of psychological trauma, and changes to immune function, reproductive behaviours and circadian rhythms (Pierce & Bekoff, 2018). All these effects were evident as Tilikum's behaviours became increasingly unnatural and indicative of psychological distress.

Thoughts of Tilikum, recently dead, surfaced when I experienced first-hand the majestic sight of a pod of killer whales off the Australian coast in 2018. Watching the almost tangible bonds between adults and young highlighted the emotional anguish inflicted on the captive Tilikum when deprived of his family from two years of age. Philosopher Martha Nussbaum's (2006) guideline for animal welfare describes a fundamental entitlement for animals to experience attachments to others, to love and care for others, and to not have those attachments warped by enforced isolation such as that imposed on Tilikum. Item 7 describes the importance of play for all sentient beings, reinforcing the need for adequate space, freedom and access to others of one's kind to achieve this. Tilikum—and the countless killer whales restrained for human entertainment before and after him—had no outlet for these basic welfare needs.

Nor did Tilikum experience the Five Freedoms of animal welfare first described in an early 1960s British government report to monitor the farming industry and subsequently expanded to animal welfare in general. (These Freedoms are discussed in more depth in Chapter 11.) Tilikum lost his freedom at two years of age, and while ready access to an adequate diet to maintain health may have been met (Freedom 1), Tilikum experienced discomfort within a human-prescribed and therefore unnatural environment (Freedom 2). His bleeding stomach ulcers showed he was not free from pain, injury and disease (Freedom 3) as he lost the freedom to express normal behaviour due to lack of adequate space, facilities and animals of his own kind when taken from his family (Freedom 4). Cumulatively, he was not free from fear and distress as his conditions and treatment promoted rather than avoided mental suffering (Freedom 5).

Analysing Tilikum's life within the context of this framework of fundamental rights lays bare the lifelong harms inflicted on him as he remained the property of humans. How this was allowed to continue for many years is not so clear cut, although the Zoological Emotional Scale can provide some insight into the shifting human perceptions of this sensitive creature.

Despite increasing research into the emotional capacity of killer whales, Tilikum's primary role in the marine parks remained financial and therefore functional. Caught up in the human entertainment industry, his positioning in the friend domain lacked credibility, being based on forced displays of affection for the audience's delight that promoted the fallacy that killer whales were attached to their carers and enjoyed repetitively entertaining humans seven days a week. The reality of Tilikum's mental anguish and disordered behaviours became evident at the third human death in 2010. While this death temporarily shifted him into the foe domain, it simultaneously provided an opportunity for change as his heart-breaking life story emerged amidst legal action, the *Blackfish* documentary and a growing awareness of the ethics of keeping killer whales and other marine mammals in captivity. Despite this, Tilikum was deemed too "psychotic" to be released and was returned to his functional entertainment status with safeguards for human physical safety rather than Tilikum's mental health, until his death in 2017.

Elephants, the terrestrial counterpart of some of the larger whales, also demonstrate a basic family unit comprising a female and her children. An older female elephant—the matriarch—her sisters, their adult daughters and all their children live together, providing the basis for shared infant care and child rearing (Safina, 2020). Status comes with age because age imbues the animal with a deep store of knowledge and wisdom that allows her family to survive. However, the ever-present threat of poachers means elephants are at risk of dying younger, resulting in psychological consequences for those left. If the matriarch is killed, family disintegration and intergenerational trauma can result.

Animal groups come together for various reasons and form what behavioural ecologists term "coalitions" (Challenger, 2021). In order to function, these groups must include some cooperative behaviours, explained over the decades through theoretical constructs such as kin selection, reciprocal altruism, direct benefits and group selection. Group social norms specific to each species provide guidance on group living and care of young. Unique examples of family and social connection proliferate in the natural world, with two common factors. These connections play some role in the survival of the individual and the species, and many are increasingly under threat from human actions.

THE IMPORTANCE OF FAMILY

After spending years focused on the lives and behaviours of wolves, Jim Dutcher and Jamie Dutcher (2013) identified a complex social unit comprising an extended family of parents, siblings, aunts, uncles and grandparents. Within this pack, there are pups who need to be cared for and educated, old wolves who need support and young adults starting to assert themselves, resulting in a dynamic and cooperative social and familial structure. Bekoff and Pierce (2009) contend that animals are capable of moral consideration as part of this group living, with sociable relationships more common among members than aggressive behaviours. Evaluations, interpretations and decision-making within their social environment are a daily necessity for social animals, but easily disrupted by the interventions of humans. Groups and packs can be broken, replaced with smaller dysfunctional units lacking the knowledge, leadership and family bonds of the original members.

Like wolves, many insects also form close familial or social bonds, living alongside each other in a community bound by kinship and rules. Social insects divide up jobs, share experiences, communicate with each other and pass on knowledge in a range of sophisticated and intricate ways (Sverdrup-Thygeson, 2019). Bees, for example, tend to their offspring with the same attention that many mammals provide their young (Tautz & Steen, 2018). Bee larvae are cleaned, fed and the pupae are kept warm by heater bees who generate heat by vibrating their powerful flight muscles. Heater bees burn up a lot of energy, and other bees will feed them by Trophallaxis—mouth-to-mouth exchange of food. Worker bees even abandon their own reproduction and wellbeing to help related individuals pass on shared genes to the next generation, often dying within a few weeks from the gruelling work of supporting the queen to pass on genetic building blocks.

Unlike bees, earwigs are often solitary insects. Despite their name, they do not have a tendency to crawl into a sleeping person's ear, preferring protected, moist environments of leaf litter and ground debris and under bark and stones from which they emerge for their nocturnal foraging. In the natural ecosystem, earwigs are sanitary engineers who clean up the environment by feeding on decaying plant material and live and dead insects.

Female earwigs care for their eggs and their young earwigs, called nymphs, an unusual trait in non-social insects and more akin to what is seen among mammals and birds. They defend their burrows from intruders, lick off fungus spores and parasites from the eggs, and then fetch food for the newly hatched nymphs, all of which increase the number of young that survive. Importantly, nurtured female earwigs mature into devoted mothers, thus increasing the survival of the overall species. Conversely, orphaned earwigs are not as effective mothers, feeding their offspring less frequently and often failing to effectively protect the eggs and nymphs from predators (Nuwer, 2016). The life of one earwig may be inconsequential to humans, but the death of a mother can have long-term implications for subsequent generations. Applying the Zoological Emotional Scale to earwigs demonstrates the divergent pathways that human perceptions can take when considering the alien-looking earwig with large forceps-like pincers at the end of a shiny black body, as shown in Figure 4.1.

Figure 4.2 represents potential interactions between earwigs and humans and the resultant effect on a person's perception of the earwig's value as a unique individual.

Figure 4.1 An earwig with large forceps-like pincers at the end of an elongated shiny brown body is a stark contrast to the more familiar mammals. Photo by Tomasz Klejdysz via Shutterstock.

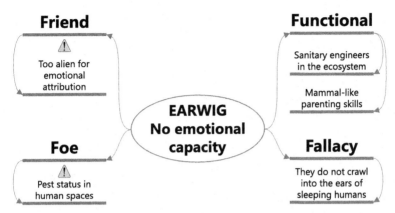

Figure 4.2 The Zoological Emotional Scale applied to the earwig indicates some reasons for the lack of emotional capacity attributed to this insect.

Any difficulties in imbuing a non-mammal with feelings may be attributable to their perceived alien appearance and lower position on traditional taxonomic hierarchies rather than accurate knowledge. Visually representing the earwig on the Zoological Emotional Scale and providing additional information on the life experiences and functional roles of an earwig can prompt the question "why do I think this way and are there other ways of perceiving this animal?" This, in turn, can encourage a person to hesitate before thoughtlessly destroying any earwigs that accidentally encroach on human-controlled spaces. However, this would also require a person to overcome learned attitudes reinforced by a lucrative pest extermination business encouraging a "them versus us" perception of these awkward creatures.

ALTRUISM AND EMPATHY

From tiny insects to huge elephants and whales, there is no shortage of anecdotal examples of individuals demonstrating altruistic behaviours where personal risk is outweighed by group or family survival advantages. Altruism, defined by Frans de Waal and Stephanie Preston (2017) as an act that promotes another's welfare even at risk or cost to oneself, has attracted considerable research among humans. One study showed that behaving altruistically activated regions of the brain that signalled pleasure and reward, suggesting that altruism may not be entirely selfless and instead a biological function shared by humans and animals alike. Once believed to require a cognitive component to weigh up pros and cons of an altruistic act and therefore belonging primarily to humans, neuroscience increasingly suggested otherwise.

Altruism without cognition became a real possibility with the discovery of mirror neurons in the 1990s (colloquially known as empathy neurons). Empathy describes the sharing of emotions, and the discovery of mirror neurons in humans confirmed that observing another's emotional state could cause a shift in the observer's emotional state to match, with no reliance on the cognitive component that humans had guarded so jealously. The recognition of another's needs through shared emotions held the potential to prompt altruistic behaviours, bringing into question humanity's self-perceived exceptionalism. Instead of empathy and altruism requiring humans to overcome their biological (animal) nature by using self-perceived superior cognitive abilities, the reverse was true. They needed to overcome their biological nature when they chose *not* to demonstrate empathy and altruism (de Waal & Preston, 2017).

Identifying mirror neurons across species, especially when relying on human models and expectations to explore the brains and nervous systems of the different, awkward animals, was challenging, but some laboratory-based research conducted among rats did detect mirror-like neurons in the anterior cingulate cortex of their brains. By inducing pain in test subject rats while allowing other rats to observe, neuroimaging confirmed that the same area that is activated in humans when experiencing, or vicariously witnessing pain was evident in both cohorts of rats (Carillo et al., 2019). While there were obvious ethical implications of inducing pain in a sentient creature for no direct benefit other than to prove a scientific point, these results also highlighted a bitter irony. Researchers had scientifically proven that rats, popular animals in biomedical research, could not only feel what was happening to them, but also to their conspecifics within the vicinity. There was no avoiding the emotional capacity within these anonymous test subjects and, by inference, the many different animals commodified for human benefit.

Fortunately, there are alternative non-invasive ways to explore empathy and altruism in a wide range of vertebrate and invertebrate animals, with ethology providing one long-standing option. Ethology encompasses the scientific study of how animals behave under natural conditions. When exploring the social phenomena of empathy and altruism within a range of animals, ethologist Marc Bekoff (2007) positioned empathy as a component of morality. As such, it required a holistic perspective of observable behaviours including cooperation, reciprocity and helping. There was no need to induce pain to observe empathy when there were abundant examples to be found in the natural world. Frans de Waal's (2016) concept of empathy and altruism similarly reiterated the importance of naturalistic observation. Describing empathy in animals as targeted helping based on understanding of the other's circumstances, de Waal cited one of the oldest documented reports in scientific literature dating back to 1954. Two dolphins kept afloat a temporarily stunned member of their pod, allowing it to breathe while they, submerged during the operation, could not. Many and varied anecdotal examples have accumulated over the years among chimpanzees, capuchins, Corvids and Eurasian Jays, and marine mammals, suggesting that empathy and altruism are the norm among animals rather than the exception.

Evidence of altruism has extended beyond humans, mammals and birds as insect behaviours gain greater attention. Reports of diseased ants who leave their colony to face death alone rather than infect the colony, ants who put their own lives at risk to free trapped members of the colony and ants who carry their wounded nest-mates back to the nest to care for their wounds suggest even the tiniest of creatures is capable of self-sacrificing behaviours for their family or social group. While these behaviours have been described as biological altruism rather than psychological altruism, the end result is the same: a selfless benefit to others that promotes social systems and group living (Birch, 2019).

Direct reciprocity, akin to altruism, but dependent on the sequential performance of costly behaviours that produce benefits, requires one individual to act beneficially towards the other after that individual has acted beneficially towards them (Freidin et al., 2017). This action was believed to be human-centric, as it required performing an action in anticipation of future benefits and weighing up the potential for cheating. Once again, observation of animals has discredited humanity's claim to uniqueness.

CASE STUDY – VAMPIRE BATS

Vampire bats, the source of considerable folklore and disgust due to their intimate connection with blood, provide an unexpected example of direct reciprocity. Adult vampire bats feed solely on blood from larger animals, including domestic livestock. They can also starve to death after 60 or 70 hours without blood, prompting other bats who have fed to regurgitate blood for their roost-mates at risk of starvation (Rocha et al., 2020). Observation of bats suggested this was based on a previously established grooming relationship rather than a direct genetic relationship, and that was a sufficient catalyst for reciprocation of regurgitated blood at a later date. The beneficial grooming relationship balanced the future potentially costly action of feeding a friend in need, as well as increasing group cohesion as a whole. Bats also demonstrated social distancing when feeling unwell, being less likely to call out to others for social interaction and thus restricting the transmission of disease.

Vampire bats—and bats in general—are increasingly associated with zoonotic pathogens, including rabies. Negative perceptions are compounded by the economic impact of vampire bats feeding on domestic farm animals as humans increasingly encroach on their territories. Selective control by veterinary services includes the capture and administration of an anticoagulant paste on the back of the captured bats, who then return to the roost. The grooming relationships fundamental to their direct reciprocity become their undoing as others ingest the paste during social grooming and subsequently die of haemorrhage.

When viewed within the Zoological Emotional Scale, the combination of blood sucking, regurgitation of blood and the risk of zoonoses—all harbingers of death in human lore—can outweigh perceptions of affective bonds through direct reciprocity. This prompts a potentially skewed positioning of this socially connected animal and minimisation of cognitive dissonance when using chemicals to destroy all inhabitants of a roost site.

These negative perceptions are not as prominent when considering the heater bee who burns up energy for the sake of the hive and is fed by other bees through trophallaxis, the mouth-to-mouth exchange of food. Or the parent bird who feeds regurgitated worms to their fledglings. Or the endangered African Wild Dog whose social structure means young pups still at the den are fed regurgitated meat in a practice no different to regurgitating blood. Despite these regurgitation habits, emotion is more easily attributed to the large-eared furry wild dog, reminiscent of the beloved domesticated dog, compared to the snarly-faced vampire bat with their leathery wings.

Figure 4.3 represents some of the possible interactions that vampire bats can have with humans, ranging from first-hand experience to fallacy, and the resultant impact on their life and death experiences. In addition to the question, "why do I think this way and are there other ways of perceiving this animal?", it is important to consider the validity of demonising a socially connected environmentally important animal who seems to be nobody's friend.

PAIR BONDING, BI-PARENTING AND THE MATERNAL BOND

Animal social and familial groups can also include specific groupings, including pair bonding, bi-parenting and the maternal bond. When people think about pair bonding,

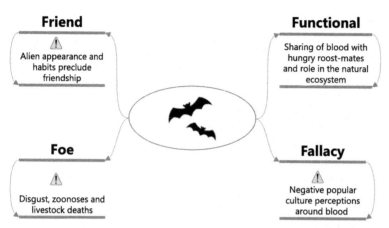

Figure 4.3 The Zoological Emotional Scale is applied to the vampire bat to identify the differing subjective perceptions and lack of emotional capacity attributed to this misunderstood animal.

inevitably their thoughts turn to human relationships and the romantic love stories that have been immortalised through books, film and social media. Some mammals may be included in the perceptions of pair bonding, again in a romanticised and sometimes unrealistic manner. The iconic image from Disney's *Lady and the Tramp* animated movie of two dogs sharing spaghetti and meatballs endows them with emotional pair bonds as strong as human love, when the reality for many domestic dogs is more in keeping with a one night stand or an orchestrated coupling on a commercial puppy farm.

Among vertebrates, monogamy and bi-parental care are highest among birds and lowest in mammals, with up to 97% of bird species forming a pair bond with a mate beyond courtship and sex. According to Professor Gisela Kaplan (2019), a prominent voice in bird behaviour, some Australian native birds become "childhood sweethearts" and may court for years before eventually "marrying". Others may "divorce" citing personality clashes and different skill levels. Parenting and bi-parenting among birds provide advantages where the young must be cared for until feathered and flight ready. Hatchling development is on a continuum ranging from super-precocial where the young are completely independent at hatching; to precocial where the young are hatched with their eyes open, covered with down and stay with their parents to be fed despite being able to leave the nest soon after hatching; to altricial where the newly hatched young are naked, blind and helpless. People perceive greater cuteness and are drawn to protect precocial species such as chickens, ducks and turkeys as evidenced by the regular social media images of law enforcement officers or citizens holding back traffic on busy highways to allow a mother duck and her ducklings to waddle across. Through association with the cute infants in a protective maternal role, the adult duck is imbued with perceived emotional capacity and awarded a level of protection not accessible to adult ducks in other situations.

For several weeks every year, Pacific Black Ducks, Mountain Ducks, Chestnut Teal Ducks, Grey Teal Ducks, Pink-eared Ducks or Wood Ducks are at risk of losing their future seasonal partners and their lives as the annual duck hunting season in an Australian state decimates their ranks. Without a clutch of ducklings lined up behind her, female ducks are no longer protected by social constructions of maternal love and devotion, instead perceived as functional commodities and legal game for human sport and entertainment. There are plenty of activists who still perceive the ducks as sentient beings capable of experiencing fear and pain, but their attempts to prevent the carnage of dead, wounded, crippled and suffering ducks can be hampered by safety and legal constraints. A duck's life thus transitions between domains of the Zoological Emotional Scale, depending on their parental status, the time of year, and whether perceived through the lens of hunter or activist. When not surrounded by cute ducklings as proof of emotional capacity, the duck in Figure 4.4 becomes a commodity of human sport.

Swans also become collateral damage as the duck hunters stomp through their territory in search of the legal, but elusive ducks. Adult swans, symbols of love and devotion on human wedding cakes and cards, mate for life and the bonded parents jointly guard and care for their eggs. Swans can become so stressed and fearful at the sound of shooting that they abandon their nests and eggs, rendering the untended eggs non-viable within hours and destroying future family life.

The *Cambridge Declaration on Consciousness*, which affirmed that humans are not unique in possessing parts of the brain complex enough to support conscious experiences, listed birds as being able to experience the positive and negative states of fear, play, anger, irritation, love, sadness and grief. Fear can increase secretion of the fight and flight hormone adrenaline, causing hyper-vigilance and enabling memory formation of the fearful event (Kaplan, 2019). The trauma associated with the gunfire and human hunters may thus become embedded in the swans' memories, to be triggered annually as the duck shooting season commences and parenting roles are once again disrupted. During the 2022 duck hunting season, a rescue volunteer described the heart-breaking outcome for a pair of devoted swan parents. Observed sitting on a nest before duck hunting season commenced, then circling the nest and calling after the

Figure 4.4 A lone duck in the wetlands is at risk of being shot during duck hunting season. Photograph supplied by the author.

shooting commenced, the volunteer noted how stressed the parents looked as they fled, leaving behind a nest of non-viable eggs (Perkins & Eddie, 2022).

According to the bi-parental care hypothesis, care of mutual offspring evolves when it offers higher reproductive success than through polygamy (Kaplan, 2019). Being monogamous is therefore a survival strategy for a species, an observation supported by reduced offspring survival in studies among a range of birds when the male parent is removed. Zebra finches, a gregarious highly social bird, are part of the 97% of birds who form monogamous lifelong pair bonds. Studies have indicated that disruptions of social interactions, such as separation from their mate or social isolation, can elicit physiological stress in these songbirds (Perez et al., 2012). In one study, male zebra finches were socially isolated from their mate, resulting in increases in plasma corticosterone concentration and an overall decline in their vocal behaviour. Just as negative experiences may cause humans to socially withdraw and seek comfort from music that matches their mood, these outgoing little birds changed their tune dramatically leaving little doubt of their physical and emotional suffering.

The ethics of inducing these strong negative emotions in a bird remains questionable and again points to research animals sometimes being manipulated simply to prove a point. Negative emotions in human studies of sadness and loneliness are often induced by viewing images, rather than physically removing them from partners and family. Another way to explore negative emotion in birds after the loss of social or family contacts is simply to watch them, although inevitably accusations of less scientific rigour compared to an artificially manipulated study may arise.

CASE STUDY – THE AUSTRALIAN LAUGHING KOOKABURRA

Watching the emotional state of an Australian Laughing Kookaburra following the loss of his life mate and three generations of offspring in a bushfire provided a moving case study outside of the laboratory. Life bonds and shared parenting are strong among the kookaburra, whose rolling chuckle and raucous chortle create the perception of unbridled happiness. The dawn chorus of laughter from high in the treetops is a feature of Australian bushland up and down the East coast. However, the laughter may cease when pair bonds are broken and families ripped apart as bushfires destroy the trees that house their preferred nesting hollows.

For more than a decade, I attended an annual workshop in the beautiful setting of Belgrave Heights in Victoria's Dandenong Ranges. The highlight was not the workshop activities, designed to further my knowledge of the human psyche and provide necessary professional development points, but rather the family of laughing kookaburras who lived on the heavily treed property. The parents, accompanied by an ever-expanding brood from current and previous years, heralded in the morning with raucous laughter that was soon echoed by rival family groups across the ranges. With the birth of each new generation, all birds in the group shared parenting duty and defended their territory. I came to recognise each member of this extended family, not just by their looks, but by their quirky personalities. Sometimes I would visit the property between workshops to watch the new hatchlings, aggressive from birth, gradually find their place in the family pecking order and eventually onto the balcony of the old house that dominated the property. Kookaburras remain in the same territory all year round, which is how the whole family ended up in the path of the 1997 bushfires that ravaged the region.

The rest of the story I learned from the property owner, who had been evacuated hours before the fires raced through the tinder-dry vegetation surrounding the house. A last-minute wind change saved the house, but not the group of trees, including the nesting hollow, that had been home to this tight-knit family for many years. The house still stood, surrounded by a blackened graveyard of trees with skeletal arms stretched out in rigor mortis and a silence befitting the graveyard that the bushland had become. The family of kookaburras had fled or died, the property owner was unsure which, until two weeks later the father kookaburra appeared alone on a blackened tree branch as if contemplating the wreckage of his home. Moving to the balcony of the property where his offspring would line up each evening, he remained there for several days, as if waiting, before disappearing and never returning again.

Natural disasters can change family structures in a flash, especially in countries such as Australia that are subject to catastrophic bushfires, drought and flood. However, a more insidious environmental impact is the relentless human encroachment on trees and greenery in suburban areas. It is not necessary to travel to distant regions to see examples of changing landscapes in a person's own local environment. This happened to a pair of magpies who for many years had roosted in one of the Eucalypt (gum) trees that dominated the streetscape of our leafy bayside town. Every year they patched up their nest in the upright fork of the tree 15 m above the ground as they prepared for a new generation to arrive. Gum trees are hardy and drought tolerant, but extended drought can result in sudden branch drop as a means of preventing death of the whole tree. Branch drop is hard to predict and so there was no way for the magpie parents to know that shortly after their eggs hatched, the nest and branch would plummet to the ground. Too young to survive out of the nest and already traumatised by the sudden descent, the fledglings were dead when humans with chainsaws arrived to remove the debris, the tree from which the branch fell and several surrounding trees representing hundreds of years of cumulative growth. Within months, the cleared area was claimed for development of human housing.

Homeless and childless, the parents remained in the area for two years without attempting to nest or breed, showing none of the cheeky inquisitiveness and melodious carolling at dawn and dusk that had endeared them to the neighbourhood. After two years, they were ready to breed again, setting up home in one of the few remaining trees several hundred metres from the original gumtree and eventually hatching out two fledglings to whom they were devoted. Figure 4.5 shows one of the new generations of magpies staring inquisitively at the camera while the parent stands protectively in front. As with the swans and kookaburras, observing animals in the wild offers meaningful evidence that disruptions in pair bonding and parenting can have long-term emotional, physical and reproductive implications.

ATTACHMENT

As with cognitive abilities, considerable research time is devoted to comparing human abilities and behaviours with those of animals as if constantly needing to establish and maintain hierarchical dominance. Birds and mammals share a phylogenetically ancient attachment system. While lactating has been found to play a role in mammal bonds of mother and infant, birds do not lactate raising speculation that attachment and care may be related to a biologically based behavioural system triggered by the fledglings' cries of

Figure 4.5 A baby magpie stares inquisitively at the camera as the parent stands protectively in front. Photograph supplied by the author.

hunger or distress (Kaplan, 2019). These cries affect both males and females in bi-parenting pairs. Prolactin, the hormone related to sex and reproduction in humans, has been found in higher levels among male and female co-parenting birds, and among Phalarope (slender-necked shorebirds) males where only the males incubate the eggs. With both humans and birds demonstrating hormone-driven parental attachment behaviours, it seems discriminatory to privilege the human's experience based on shared language to articulate their feelings and disregard the bird's attachment to their fledgling as instinct.

While long-term pair bonding and bi-parental or group care similar to human systems are common among many birds, it is more limited among mammals. One exception is the diminutive meerkat, found in the arid Kalahari where scarcity of food and few places to hide from predators has led them to live in large family groups of up to 50 individuals for survival. One dominant male and female breed, producing up to three litters and 20 pups a year, and their pups are fed and raised by the group (Clutton-Brock, 2006). So popular has the meerkat become after the Disney animated movie *Lion King* (1994) and Animal Planet's documentary *Meerkat Manor* (2005–2008), it is easy to ascribe emotion to these tiny creatures with emotive faces and the capacity to stand on two legs and turn as one in the same direction (Figure 4.6). *Meerkat Manor*, which focused on a group of meerkats in the Kalahari desert of Africa made viewers feel as if they knew the animals, all of whom had names and their lives narrated as if in a soap opera.

It becomes easy to forget that meerkats are important members of the natural environment, maintaining ecological balance by curbing pest infestations and providing food for predators such as jackals and eagles. Instead, their cuteness factor has resulted in meerkats being sold as pets in some countries, despite them being social animals whose complex needs cannot be met in captivity away from their family structure. They can develop behavioural and physical problems which see them make use of their long, strong claws. It seems that even a cute emotive animal such as the meerkat can be subject to negative consequences. Apart from being adopted as pets and removed from

Figure 4.6 A meerkat in captivity sits upright on a rock and gazes forward. Photograph supplied by
the author.

their important social structures, meerkats have also been gassed in their burrows for
potentially carrying rabies when humans encroach on their territories. The Zoological Emotional Scale in Figure 4.7 provides a visual representation of the complex life
experiences of the meerkat in a human-dominated world.

Given the meerkat's emotive expression and cuteness factor, it can be easier to attribute family devotion and parenting skills to them compared to a notorious reptile with
a killer reputation. Parenting comes in all forms, with many behaviours that may seem
alien to humans but are nevertheless devised for protection of the young. While African
Nile crocodiles do not display the nurturing bi-parental behaviours seen in birds, the
male may still hang out near the nest while the female guards (Benyus, 2014). When
it is time to hatch, babies "bark" loudly, prompting mother to make sure they have all
managed to break free from the leathery lining of their eggs and assist her new babies
to scramble out of the packed-down soil of their nest. All the babies then climb into her
open mouth for transport to the water's edge for release. Should a hatchling be in trouble, it broadcasts a distress call causing all hatchlings in the area to similarly cry in distress,
reminiscent of emotional contagion among a room full of infant humans. The mother
lurks close by for the first 12 weeks of the infants' life, an example of maternal care
crocodile style. This may be based on an instinct to preserve her young, but the effect is
just the same as if she were a doting parent displaying mammal or bird-like behaviours.

MATERNAL BONDS

Compared to the less common bi-parenting behaviours among mammals, maternal
bonds are widespread. An early maternal separation study by American psychologist
Harry Harlow (1958) poignantly demonstrated the impact of separating rhesus monkeys from their mothers. Removed from their birth mothers at a young age, infants
were given surrogate mothers made of wire or terrycloth, both of which provided milk

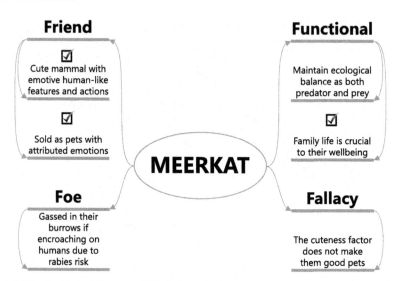

Figure 4.7 The Zoological Emotional Scale is applied to the Meerkat to identify where emotional capacity is attributed to this family-oriented animal.

at different times. The infants consistently sought comfort from the cloth surrogate, irrespective of whether it was offering milk, and cried and screamed when denied this access. Based on this ethically questionable study, Harlow concluded that maternal bonds and comfort were even stronger than the need for food at times.

Oxytocin, the hormone associated with maternal love, is evident in human attachment relationships, as well as some animal attachment relationships within and at times between species. For humans in contemporary societies, companion animals may be their only contact with live animals and increasingly the attachment bond between human and companion animal is normalised as the animal achieves kin status in some human families (for example, see research by Irvine & Cilia, 2017 and Laurent-Simpson, 2021). When humans and dogs gaze at each other in mutual adoration, the same neurological mechanisms that bond human mother to child are activated in what has been described as an oxytocin-mediated positive loop (Nagaswa et al., 2015). Sustained mutual gaze releases oxytocin in both parties, strengthening their bond and leaving them at risk of negative emotions should the bond be broken.

Other domestic animals remain abstract concepts for many people, disconnected from the packaged meats in supermarkets and on their plate. To acknowledge that the source animals of these products are capable of forming attachments and maternal bonds through the same neurological mechanisms that bond human mother to child risks cognitive dissonance and avoidance strategies. The disconnection of maternal bonds and end product is particularly evident in the production of milk and other dairy products for human consumption. This is the story of Clarabelle, a mother cow who was finally permitted to develop a maternal bond with her infant in the safety of Edgar's Mission, a sanctuary for rescued farmed animals in Victoria, Australia.

Clarabelle started life as a dairy cow who was regularly impregnated, only to have her calves taken away at birth so that she could continue producing milk for humans. After arriving at Edgar's Mission in 2014, Clarabelle gave birth to Valentine, and for the first time in her life was able to nurse her baby and maintain the bond between them. Clarabelle's story is reproduced with permission from Edgar's Mission.

Clarabelle and Valentine, as Told by Edgar's Mission

Clarabelle's behaviour told us something was amiss. Although her baby was not expected for another week, something was odd. A firm favourite to be first to feeding, on this occasion Clarabelle was not. Apprehensively, she walked up the paddock, every now and then casting a quick glance behind. A rather engorged sole teat was our first hint this eight-year-old Jersey cow had something to hide.

But where? With a clear view of the paddock, no little calf lay. But an enchanted forest that claimed the tiniest section of the rear part of this field offered an answer. With sleuth-like precision, we made our way towards it but nothing looked suspicious. Nothing until we almost stood on a tiny bundle of brownness, ever so carefully hidden in the tall grass and camouflaged by fallen logs. We believe there are few things on this earth more precious and innocent than a baby calf. This little heifer, as she lazily blinked at the world with big bug eyes could surely have melted a polar ice cap. But this was no newborn calf; fully clean and dry was she, along with her umbilical cord—no afterbirth in sight.

Many who know cows will vouch they have an ancient knowing, a wisdom beyond their form. They are like elephants in their memories. Renowned animal behaviourist and cattle expert, Dr. Temple Grandin, is credited with saying that the fear memories of cattle can never be deleted. This fact was borne out by the kindly dairy farm worker (from where Clarabelle came) who told us that the cows remembered which vehicle came and took their baby away shortly after birth. On subsequent occasions when farm vehicles would drive past they would behave no different, no different that is until the one vehicle that took their baby would return. At this point, the cow would become nervous, anxious and edgy, looking for the baby she would never see again.

Bless this sweet girl, who, having had each successive calf taken from her shortly after it was born, was determined this one would be "hers". So stashing her baby in the forest she walked up the paddock as if nothing had changed. But everything in fact had, forever more things will be the way they always should have been and no one will ever take Clarabelle's baby away.

While Clarabelle hid her baby for the most pure, honest and loving reasons, the dairy industry has long hidden the fate of baby calves for reasons of profit. Like many we grew up with the romantic notion that dairy products were wholesome and good, and indeed they are. If, of course, you are a baby cow. Cows do not produce milk simply by virtue of being a cow; they are mammals and will only produce milk for their offspring. Once born, the babies are soon taken away from their mothers and the milk they intended for their baby is harvested for human consumption. The males, who will never produce milk, are killed, many in the first week of their life.

Females can become herd replacement animals and some are even sent to China to meet a growing demand for dairy. Small or non-commercial heifers meet the same fate as the males. If you love dairy products and don't believe you can ever find an alternative, please remember this, mother cows love their babies many times more.

Reproduced with permission from Edgar's Mission.
Images provided by Edgar's Mission show Clarabelle with a very young Valentine (Figure 4.8), and Clarabelle retaining the bond with the grown-up Valentine (Figure 4.9).

Figure 4.8 Clarabelle leans over the newborn Valentine. Image supplied by Edgar's Mission and reproduced with their permission.

Figure 4.9 Clarabelle licks the grown-up Valentine. Image supplied by Edgar's Mission and reproduced with their permission.

KANGAROO CONTRADICTIONS

Attitudes to orphaned marsupials, including kangaroos, vary around Australia. Marsupials such as kangaroos, wallabies and wombats give birth to "joeys", tiny, hairless, blind babies with only partially formed limbs. Carried in their mother's pouch for months after birth, they remain helpless should their mother be one of the tens of thousands of marsupials hit by motor vehicles each year. Although they can survive for up to five days in their dead mother's pouch, they eventually face death by starvation or cold. Wildlife organisations provide information for volunteers on how to check a dead mother's pouch for living offspring in need of care. These dedicated volunteers spray paint a large cross on the body of animals they have checked as they desperately try to save the helpless infants hidden in their dead mothers' pouches.

Contrast that with the culling of kangaroos when their numbers expand and they threaten the pastoral industry. National Codes of Practice for the shooting of kangaroos and wallabies for commercial or non-commercial purposes specify that if a shooter kills a female with pouch young, they must also kill the joey. This is usually done by decapitation, a blow to the head or shooting, depending on the size of the joey. If the shooter does not kill them, dependent young are at risk of a slow, painful death (RSPCA, 2020). While the deaths of kangaroo mothers in both cases are human instigated, there is a marked difference in the treatment of their offspring depending on how the mother kangaroo is perceived. As a foe to the pastoral industry, the joey will be killed. As a friend to the volunteers, the joey, like the one in Figure 4.10, will be rescued and rehabilitated.

Figure 4.10 A joey sits alone behind an old log. Photograph supplied by the author.

IMPLICATIONS

The mechanisms that bring together and maintain social groups, families, pair bonds and parents may vary, but the implications for severing those groupings can be far-reaching for the individual and the species. From killer whale orphans who become psychotic, to orphan earwigs who struggle to mother their own nymphs, to Clarabelle the cow who hid the baby she was so determined to keep, the evidence is undeniable. Many vertebrate and invertebrate animals are reliant on maintaining intact social, familial or parental bonds. Whether these responses are emotional or instinctual, or a mix of both, is irrelevant when considering the right of all animals to experience the fourth and fifth domains of freedom: freedom to express normal behaviour through the provision of adequate space, facilities and animals of their own kind and family; and freedom from fear and distress through the provision of conditions and treatment that avoid mental suffering.

Research has largely dispelled controversy over whether vertebrate and many invertebrate animals have the capacity to experience positive and negative feelings, demonstrate altruistic and empathic behaviours and form strong social, familial and parental bonds. However, humans retain the power to sever the parent/child and family bond of so many animals, to the detriment of the infant, the parent and the species. As animal culling, poaching, legalised hunting, illegal trafficking, road accidents and destruction of habitats all impact animal family life, the Zoological Emotional Scale can provide some insight into the contradictions in human perceptions that privilege one animal over another. Where these contradictions give rise to cognitive dissonance, a person may continue drawing on justifications or denial, or move into the pre-contemplation, contemplation and preparation stages of change.

A person's relationship with animals will also direct their ponderings as the veterinary professional in a public clinic will have different interactions and perceptions of animals compared to the veterinarian attached to a research facility, just as the animal rescue worker will have different interactions and perceptions compared to the animal slaughterhouse worker. This makes the cycle of change a very personal journey drawing on a person's life circumstances as well as their desire and capacity to make a difference.

Acknowledging parental, partner, family and social bonds among animal groups raises questions as to possible outcomes of grief, loss and trauma when these bonds are broken. This is explored in Chapter 6. Because emotions and grief can never really be known in another living being, especially animals, anthropomorphism has become a common mechanism of interpretation. The benefits and pitfalls of humans imposing perceived feelings on another sentient being are discussed in the next chapter.

REFERENCES

Bekoff, M. (2007). *The emotional lives of animals: A leading scientist explores animal joy, sorrow, and empathy–and why they matter.* New World Library.

Bekoff, M. & Pierce, J. (2009). *Wild justice: The moral lives of animals.* The University of Chicago Press.

Benyus, J.M. (2014). *The secret language of animals.* Black Dog & Leventhal Publishers, Inc.

Birch, J. (2019). Altruistic deception. *Studies in History and Philosophy of Science Part C, 74,* 27–33. https://doi.org/10.1016/j.shpsc.2019.01.004

Carillo, M., Han, Yinging, Migliorati, F., Liu, M., Gazzola, V. & Keysers, C. (2019). Emotional mirror neurons in the rat's Anterior Cingulate Cortex. *Current Biology, 29*(8), 1301–1312.e6. https://doi.org/10.1016/j.cub.2019.03.024

Center for Whale Research. (2023). *About Orcas.* https://www.whaleresearch.com/aboutorcas

Challenger, M. (2021). *How to be animal. A new history of what it means to be human.* Canongate Books Ltd.

Clutton-Brock, T. (2006). Competition and cooperation in meerkat reproduction. *Nature,* 444be, xi. https://doi.org/10.1038/7122xia

Cowperthwaite, G. (2013). *Blackfish.* Magnolia Pictures.

De Waal, F.B.M. (2016). *Are we smart enough to know how smart animals are?* W.W. Norton & Company Inc.

de Waal, F.B.M. & Preston, S.D. (2017). Mammalian empathy: Behavioural manifestations and neural basis. *Nature Reviews Neuroscience, 18*(8), 498–509. https://doi.org/10.1038/nrn.2017.72

Dutcher, J. & Dutcher, J. (2013). *The hidden life of wolves.* National Geographic Society.

Freidin, E., Carballo, F. & Bentosela, M. (2017). Direct reciprocity in animals: The roles of bonding and affective processes. *International Journal of Psychology, 52*(2), 163–170. https://doi.org/10.1002/ijop.12215

Harlow, H.F. (1958). The nature of love. *American Psychologist, 13,* 673–685.

Irvine, L. & Cilia, L. (2017). More-than-human families: Pets, people and practices in multi species households. *Sociology Compass, 11,* e12455. https://doi.org/10.1111/soc4.12455

Kaplan, G. (2019). *Bird bonds. Sex, mate-choice and cognition in Australian native birds.* Pan Macmillan Australia Pty. Ltd.

Laurent-Simpson, A. (2021). *Just like family: How companion animals joined the household.* NYU Press.

Nagaswa, M., Mitsul, S., En, S., Ohtani, N., Ohta, M., Sakuma, Y., Onaka, T., Mogi, K. & Kikusui, T. (2015). Oxytocin-gaze positive look and the coevolution of human-dog bonds. *Science, 348*(6232), 333–336. https://doi.org/10.1126/science.1261022

Nussbaum, M.C. (2006). *Frontiers of justice: Disability, nationality, species membership.* Belknap Press of Harvard University Press.

Nuwer, R. (2016, April 1). Orphaned Bugs Make Bum Parents: Researchers show earwigs pass trauma to their offspring. *Scientific American.* https://www.scientificamerican.com/article/orphaned-bugs-make-bum-parents/

Perez, E.C., Elie, J.E., Soulage, C.O., Soula, H.A., Matheyon, N. & Vignal, C. (2012). The acoustic expression of stress in a songbird: Does corticosterone drive isolation-induced modifications of zebra finch calls? *Hormones and Behavior, 61,* 573–581. https://doi.org/10.1016/j.yhbeh.2012.02.004

Perkins, M. & Eddie, R. (2022, May 19). Swans flee nests as duck hunting season starts. *The Age.* https://go.gale.com/ps/retrieve.do?tabID=T004&resultListType=RESULT_LIST&searchResultsType=SingleTab&hitCount=535&searchType=AdvancedSearchForm&currentPosition=1&docId=_GALE%7CA697219019&docType=Article&sort=Relevance&contentSegment=ZNEW-FullText&prodId=STND&pageNum=1&contentSet=GALE%7CA697219019&searchId=R1&userGroupName=unimelb&inPS=true

Pierce, J. & Bekoff, M. (2018). A postzoo future: Why welfare fails animals in zoos. *Journal of Applied Animal Welfare Science, 21,* 43–48. https://doi.org/10.1080/10888705.2018.1513838

RSPCA Australia. (2020). *What happens to joeys when female kangaroos are shot?* https://kb.rspca.org.au/knowledge-base/what-happens-to-joeys-when-female-kangaroos-are-shot/

Rocha, F., Ulloa-Stanojlovic, F.M., Rabaquim, V.C.V., Fadil, P., Pompei, J.C., Brandão, P.E. & Dias, R.A. (2020). Relations between topography, feeding sites, and foraging behavior of the vampire bat, *Desmodus rotundus. Journal of Mammalogy, 101*(1), 164–171. https://doi.org/10.1093/jmammal/gyz177

Safina, C. (2020). *Becoming wild. How animals learn to be animals.* Oneworld Publications.

Sverdrup-Thygeson, A. (2019). *Extraordinary insects.* Mudlark.

Tautz, J. & Steen, D. (2018). *The honey factory: Inside the ingenious world of bees.* Black Inc.

ANTHROPOMORPHISM OR EMOTION?

Humans have a tendency to project hypothetical feelings onto nonhuman animals. This can become a natural and spontaneous way to relate to all animals, both familiar and unfamiliar, despite controversy over the flaws and benefits of this method of interpretation. It is no wonder that the concepts of anthropomorphism and anthropocentrism permeate human–animal research and literature. Anthropomorphism, from the Greek words Anthropos (human) and morph (form) and defined as characterisation of nonhuman behaviour or objects in human terms, is an enduring and seemingly irresistible human inclination (Wynne & Udell, 2013). In his book *The Ape and the Sushi Master* (2001), primatologist and ethologist Frans de Waal (2001) commented that closeness to animals instils a desire to understand what goes on in their heads, while simultaneously realising that this is impossible beyond an approximation based on human behaviour. Anthropomorphism may thus provide a starting point, laden with both risks and gains.

Recognising that animals are still animals but persisting in viewing them from a sometimes skewed human perspective can result in anthropocentrism, literally defined as human-centred. This viewpoint is reinforced by the use of human language as the medium when projecting hypothetical assumptions on to an animal, whether it be the companion animal's human speaking on their behalf (often in a baby voice reminiscent of parentese), or through literature that attempts to afford real-world nonhuman animals their own "voice". Aaron M. Moe (2016), writer and researcher in animal rhetoric, animal studies and ecocriticism, wrote

> Other disciplines [...] have helped society and culture undergo a colossal paradigm shift with how we understand, respond to, and interact with nonhuman animals [...] literary studies may seem to be one of the last places that explores the lived realities of actual, biological animals given its reputation for turning animals into anthropomorphised tropes.
> (p. 133)

While human language may provide one means to look within the animal, fields such as communication studies and biosemiotics make it increasingly clear that human language is not the only form of communication. Animal behaviourists Jack Bradbury and Sandra Vehrencamp's (2011) book *Principles of Animal Communication*, for example, identifies numerous forms of communication, including movement, sound, visual

DOI: 10.1201/9781003298489-6

signals and chemical signals. Anthropologist Don Kulick's (2017) work on human–animal communication specifies that there remains a need to:

> [...] explore and extend the grounds for respectful engagement with animals in ways that do not either reduce animals to anthropomorphic projections, or claim them to be fundamentally unknowable aliens whom we can continue to exploit because we can never know what, or even if, they think.

> (p. 359)

Similarly, Italian philosopher Emanuela Cenami Spada (1997) described anthropomorphism as an inevitable risk necessitated by having to refer to human experience to be able to formulate questions about animal experience. Animals, he stated, present an embarrassing problem because humans are themselves animals, but continue to perceive themselves as distinct and unique.

Anthropomorphism necessarily plays a part in the Zoological Emotional Scale, given that the scale is premised on human perceptions of animal emotional capacity and therefore a social construct. Anthropomorphism comes in different forms, ranging from the anthropocentric question "How would I feel in this situation?" to the animal-centric question "What is it like to be a XXXX?" (de Waal, 2001, p. 77). Drawing on Marc Bekoff's (1997) contention that being anthropomorphic does not necessarily ignore the animal's perspective, instead making accessible the thoughts and feelings of another sentient being during a shared experience, this chapter examines the good, the bad and the ambiguity of anthropomorphism.

In the 1940s, decades before it would become acceptable, comparative psychologist Donald Hebb (1946) balanced his non-attributive descriptions of captive chimpanzees' behaviours with the keepers' anthropomorphic perceptions of their feelings and actions. The result was a psychological account of the chimpanzees' behavioural patterns, which included attributions of emotions such as fear and nervousness from those who related to the animals in a different way to Hebb, as a researcher. While the published results failed to mention the anthropomorphic origins of some of the data so as to maintain current scientific standards for credibility, Hebbs' decision to work with, rather than categorically eliminate, anthropomorphism was ground-breaking.

Anthropomorphism can be instrumental in raising human concern and empathy for the wellbeing of some animals, but it can also engender misunderstanding of other animals. In 2006, blogger Ben Grossblat drew attention to the many advertising images of animals who seemed delighted to be killed and eaten. Describing these animal images as "suicide food", Grossblat's (2011) blog highlighted cartoons of pigs, chickens and lobsters rejoicing as they offered their bodies for human consumption in fast food outlets, restaurants and homes. Similarly, there can be unrealistic anthropomorphic depictions of farmed animals having emotionally fulfilling lives as they suffer relentless milking or egg-laying schedules or await slaughter. For urban dwelling children and adults with limited or no contact with farmed animals, these perceptions may allay the potential cognitive dissonance if confronted with the realities of life and death within the animal agriculture industry.

Louie the fly, the anthropomorphised household fly mentioned in Chapter 2, features as a lovable, cheeky character who is comfortable in his germ-ridden status and philosophically accepting of the need for his death by Mortein fly spray. Born in 1957 and endowed with the iconic jingle about his exploits in 1962, Louie and his insect friends cavort happily to the words that describe his journey from a rubbish tip to the human household, spreading disease as he goes. Louie became immortalised with the addition of the jingle "Louie the Fly" to the National Film and Sound Archive of Australia (NFSA) in 2017. The cognitive dissonance triggered by the juxtaposition of a street-smart, roguish anthropomorphised fly who was the apple of his doting mother's eye with Mortein, an instrument of death was easily allayed by the threat of disease. In later years, Louie was joined by other insects, including spiders, mosquitoes and cockroaches, ostensibly showing the versatility of Mortein to kill all insects with no differentiation, but further consolidating the belief that the lives of insects are subject to human control with no negative implications for their natural ecological systems. Figure 5.1 shows the outcome of indiscriminate use of insect spray on both flies and all other insects within the vicinity, demonstrating that the reality of killing insects is different to the anthropomorphised jingle about Louie the fly.

ANIMATED ANTHROPOMORPHISM

While Louie the fly and friends were anthropomorphised as a marketing tool and suicide animals anthropomorphised to allay cognitive dissonance, the familiar companion animals are also anthropomorphised on a daily basis. This positions their wellbeing and perceived happiness in the hands of human caretakers as they attempt to interpret their animal's needs. When exploring *What Dogs See, Smell and Know* (2010), author and dog cognition specialist Alexandra Horowitz commented that trying to understand a dog's perspective was like being an anthropologist in a foreign land. Humans will never achieve a perfect translation of every wag and woof, she continued, but that is no

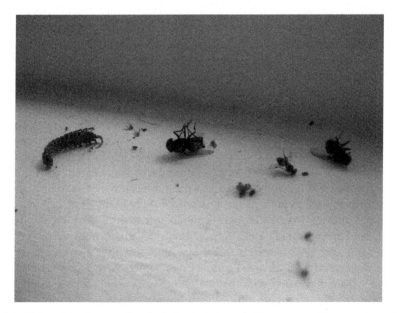

Figure 5.1 A host of dead insects after the indiscriminate use of insect spray. Photograph supplied by the author.

reason to stop looking, while simultaneously avoiding the tendency to view every moment of a dog's life through a human lens. This included the increasingly popular trend to de-animalise the dog with human artefacts such as clothes and birthday parties.

Media depictions of anthropomorphised animals in animated films often depict them within human-like social systems, irrespective of whether they are companion animals, farmed animals, wild animals or invertebrate animals. Animated films, in particular, have been described as powerful social artefacts for the socialisation of children (Pandey, 2004), risking the generalisation of an animal's anthropomorphised characteristics to real animals. This, in turn, can create unrealistic future expectations where children have limited contact with the real-world animals. Walt Disney and the Disney Studios recognised and utilised the power of animated animal characters by creating a reality where moral messages were conveyed through representations of speaking animals, with the unfortunate side effect of merging human qualities with the traits of animals and promoting unattainable expectations and treatment of animals.

Universal Picture's 2016 animated movie *The Secret Life of Pets* provides one such example. This lively film transforms the boring, isolated lives of dogs, cats, birds and hamsters locked in Manhattan apartments while their humans are at work into high adventure. Any abandoned and unwanted animals are similarly portrayed as enjoying life amidst exotic gangland comradery in the sewers. Depicting apartment animals as happy and well-entertained in their human environments denies the behavioural issues such as separation anxiety that may arise when animals with emotional and social needs are confined alone. This box office hit movie subtly reinforces social norms around domestic animals as property to be put away when not needed within the isolation of small apartments and ascribes humour to any resultant behavioural issues that were then remedied with human products (for example, an overweight cat who raided the refrigerator and a budgerigar who escaped his cage to enjoy action movies while flying into the air stream of a fan). The negative consequences of these misrepresentations became reality when some companion animals—obtained during the global lockdowns of the COVID-19 pandemic to allay human isolation—subsequently developed behavioural problems as human activities resumed and they were left to emotionally fend for themselves.

Stories with anthropomorphised insects also play a role in potentially misrepresenting the voices of animals. *The Ant and the Grasshopper*, one of Aesop's fables, provides a false narrative of stereotypical insect life being the same across all species as a way of modelling theories of human social interactions. The industrious ants are depicted busily storing food for winter while the happy-go-lucky grasshopper is singing and playing all summer. This supports the human social narrative of the benefits and necessity of planning and saving, but totally disregards any biological adaptations of the grasshopper that sees them surviving successfully in different ways.

Animated movies with anthropomorphised insects date back to the 1930s (for example, the Disney Studio's 1932 *Bugs in Love*), but three animated movies released between 1998 and 2007 clearly position hive insects within a culture more akin to human individualism. Dreamworks' *Antz* (1998) depicts Z, the worker ant, struggling to reconcile his individuality with the communal work ethic of the ant colony. Pixar's *A Bug's Life* (1998) tells the story of Flik an inventive ant who destroyed his colony's food supply and sought help from a band of high-flying circus insects. Paramount Picture's *Bee Movie* (2007) introduces Barry the Bee who was determined to obtain justice for bees by suing humanity for theft of honey. In each case, the collectivist culture of the colony insects was re-storied as a barrier to the individual who strives to be different and escape their pre-ordained task, a story-line reminiscent of many human struggles.

ANTHROPOMORPHISM IN LITERATURE

The graphic novel *Letters from Animals to Those Who Think They're Just Beasts* (Brrémaud, & Rigano, 2021), based on the essays of wildlife conservationist Allain Bougrain-Dubourg, ostensibly seeks to give animals their own voice to make themselves heard while simultaneously providing a unique focus on animal protection. Bougrain-Dubourg's introduction to the graphic novel questions how to find the right words and how to translate the emotions and requests of the animal kingdom because, he muses, the act of offering animals a human voice to express their grievances necessarily gives in to anthropocentrism. Bougrain-Dubourg does not directly answer this question, rather pointing out that if any human behaviour hurts animals it should be condemned. He expands on this: "I've fought for a long time, together with others, to defend them [animals] without achieving the desired results. I thought that by giving them the chance to speak, they would make a better case for themselves!" (p. 92).

However, reading through some of the proposed "better cases" in the graphic novel, it becomes difficult to locate the animal's own voice in some of the interpretations provided. For example, the pig writes: "In the wild, we can live to 25 years old. In industrial farms, that's reduced to just a few months. And even if after considering all these horrors, you still don't want to give up eating pork, at least support farms that use the most sympathetic methods […] Look for labels that say 'organic', 'open air', or better yet, 'free range'". This sage advice may buffer human feelings of cognitive dissonance, but it also overlooks the bitter irony of an intelligent, emotional creature offering suggestions on their own demise for those humans who are unable to give up eating pork.

The wolf also requests discriminate killing of their kind, while completely ignoring the events that saw humans encroach on their homeland: "Many of the wolves you kill are from packs that don't cause any problems. Be fair and focus on the ones that actually attack herds. Let the innocent ones live". Like the pig, the wolf seeks to allay human cognitive dissonance by acknowledging that some of their kind deserve to die when affecting human livelihoods, an anthropocentric assumption spoken through the voice of the animal.

Despite some potentially inaccurate anthropomorphic portrayals, the uniqueness and vulnerability of animal culture is increasingly gaining an animal-centric voice. Ecologist Carl Safina (2020) commented that animal culture has remained hidden and is best understood by going deep into nature and looking at individual animals in their own communities. While this is impossible for many humans and would undoubtedly destroy the object of their observations through further human encroachment, it is possible to differentiate between anthropomorphism that furthers the human agenda and anthropomorphism that attempts to make sense of what an animal may be feeling. Perpetuating human values through inaccurately anthropomorphising animal culture de-stabilises the rights of animals to their own ways of thinking, feeling and experiencing life. It also reinforces and perpetuates the belief that the only way to understand and present animal stories is by articulating them in human language.

THE BENEFITS OF ANTHROPOMORPHISM

Marc Bekoff (2007) stated what seems to be obvious when he commented that by studying animals, humans can only describe and explain their behaviour using words that make sense from a human-centred point of view. However, he added the important qualification: "Just because I say a dog is happy or jealous, this doesn't mean he's happy or jealous as humans are, or for that matter as other dogs" (p. 123). Carl Safina makes a similar point when asking the logical question: if we recognise that an animal is hungry or thirsty, why cannot we also recognise that they are happy or joyful when acting that way? (Safina, 2015).

Acknowledging that anthropomorphism is a linguistic mechanism to make the thoughts and feelings of animals more accessible, rather than more correct, can make it a valuable tool in animal studies, as Hebb discovered decades earlier. Despite the indisputable similarities and functional equivalences between human and nonhuman animal nervous systems, it is impossible to get inside the head of an animal to check thoughts and feelings. This leaves interpretation of overt behaviours as a necessary option, introducing the risk of basing outcomes on behaviourism, a dominant theoretical framework often used to understand animal intelligence rather than emotional capacity (Safina, 2015). The myth that animals are driven by instinct not emotions is thus perpetuated, reinforcing anthropocentrism and the superiority of humans in the emotional domain while overlooking that humans are themselves animals and therefore subject to the same conclusions being drawn.

During a series of interviews I conducted to explore the human–animal relationship, it became evident that anthropomorphism based on familiarity and respect was a powerful mechanism to reinforce a meaningful relationship. When speaking of the chickens and ducks on their rural property, one person described how the birds were free to roam wherever they wanted to, including through the house. She commented: "They've got free run of the place and they respect that. I'll say, don't poop over there, if you need to go, go outside, and they listen and poop outside".

Later in the discussion she clarified that all the animals had their own personalities which sometimes made her question her thoughts. She cited the example of telling the chickens that they looked sad one day, then wondering whether that was, in fact, her own anthropomorphic projection. Reminiscent of Bekoff's earlier comments, this person's assumption that the chickens were sad and that toileting outside was respectful did not mean they were sad or respectful in the human sense. It did, however, provide a powerful linguistic tool to strengthen the bond between this person and the animals with whom she had chosen to share her home and her life.

Another person described his relationship with the chickens, roosters and ducks on his rural property. Speaking of the rooster, he laughingly admitted that if he had occasion to be strict with the bird, later that evening the rooster would come into the house and perch on his armchair and watch television with him rather than going to the hen house. He added: "He's saying, I love you, and I'm really sorry". Similarly, based on the altered behaviours of one of the ducks, the person translated this into the duck's belligerence at being left behind when he had to travel to the city overnight. He described the duck turning his back on him, which was completely out of character, as soon as he saw the car being packed.

While impossible to know for sure if these anthropomorphised behaviours and feelings were correct outside of a human context, it provided recognition that these birds were capable of experiencing meaningful emotions and strengthened the connection between humans and animals. One person did stress to me that even though he was treating the birds in the same way that he might treat a human, he was very aware that they were not people and still retained their own animal wants and needs to which he was not always privy.

CASE STUDY – THE MEANING OF RABBITS

Rabbits have been imbued with a range of meanings, many of them based on animated caricatures and edible replicas of this small resilient mammal. The perceived emotional capacity of the rabbit is therefore subject to a range of interpretations and collective experiences starting in childhood. For urban children, rabbits may be cute, cuddly low-maintenance pets capable of giving and receiving affection. This perception is reinforced by lively cartoon characters including Beatrix Potters' *Peter Rabbit*, Looney Tunes' *Bugs Bunny*, *Alice in Wonderland's* white rabbit and *The Secret Life of Pets'* Snowball who foregoes his roguish captain status for ownership by a human child. Alternatively, in some cultures, the rabbit comprises part of the Christian Easter celebrations as both a fluffy mystical creature and an edible chocolate replica.

Moving away from the positive depictions, the rabbit may also be positioned as a voracious eater and breeder, capable of destroying land and crops and meriting targeted destruction by hunting, trapping and biological means. Rabbits arrived in Australia with the First Fleet that brought British colonists, convicts and domestic animals to the new colony in 1788 (West, 2018). Deliberately released in Victoria in 1859, their rate of spread throughout Australia was the fastest of any colonising mammal and resulted in irreversible land degradation and other costly damage. Biological attempts to eradicate the perceived alien intruder, including the cruel disease Myxomatosis in 1950 followed by the Calicivirus in 1996, saw Calicivirus quickly spread into suburban domestic rabbits, necessitating regular vaccinations for those companion animal rabbits who had achieved friend status.

Rabbits have been further vilified in the human-evoked rivalry with a long-eared endangered Australian marsupial called the Bilby. In a bid to save the Bilby, the cute marsupial has been positioned as a replacement for the traditional chocolate Easter bunny in Australia. *Easter Bilby*, an anthropomorphised fictional children's story supported by the Anti-Rabbit Research Foundation of Australia, adopts the rabbit's own voice to criticise rabbit behaviours:

> Easter Bunny says, 'Bilby, I want you to have my job. You know about sharing and taking care. I think Australia should have an Easter Bilby. We rabbits have become too greedy and careless. Rabbits must learn from Bilbies and other bush creatures.'
>
> (Garnett & Kessing, 1994)

Nevertheless, Easter Rabbits in all shapes and sizes remain dominant in the retail stores each year with the tiny bilby overshadowed by the variety of rabbits to tempt shoppers, as Figure 5.2 indicates.

Anthropomorphism can be problematic when it privileges inferred human-like traits while ignoring factual evidence. Greed and carelessness are human characteristics. Rabbits eat to survive rather than through greed, while carelessness is a luxury

Figure 5.2 Four different versions of the rabbit as a popular and traditional symbol of Easter compared to the less plentiful chocolate Bilby at the front. Photograph supplied by the author.

few wild animals can afford if they are to avoid predators. Imposing subjective human attributes to animals can have opposing effects. Firstly, seeing an animal as human-like with intention, desire and regret creates a being that deserves concern for its wellbeing (Epley et al., 2008). Alternatively, attributing negative intentions and emotions to the animal dehumanises them, thus relieving the observer of any concern for their wellbeing and the resultant cognitive dissonance at their treatment.

As observed with the rat in Chapter 2, visually representing the rabbit on the Zoological Emotional Scale draws attention to the many conflicting roles that this furry little mammal holds within a human-dominated world. Rabbits occupy around 70% of the Australian mainland and their treatment is very much dependent on their geographic location and their human-assigned roles. Rabbits can be cute and cuddly companion animals with emotive faces, twitching noses, demonstrated attachment behaviours and capacity for emotional displays of enjoyment, play, fear and distress. As functional research subjects, these capacities cannot be acknowledged if the rabbit is to fulfil its role as an anonymous emotionless commodity worthy of humane treatment but still positioned below human needs. Both the foe and fallacy domains tip the scale to lack of emotional capacity as rabbits threaten native wildlife and human property, live up to the derogatory saying "breed like rabbits", and eat voraciously (but not greedily) to survive in the land to which humans introduced them.

Figure 5.3 demonstrates these competing positions of the rabbit as a friend with emotional capacity and therefore to be valued as a unique individual; a functional creature to be valued for what it provides humanity; a foe with introduced pest status in human-dominated spaces and therefore a threat to be destroyed; or a misrepresented creature based on fallacy. First-hand experience plays a role in the friend, foe and functional domains, although for very different reasons, while learned attitudes are influential in the foe and fallacy domains. When applying the Zoological Emotional

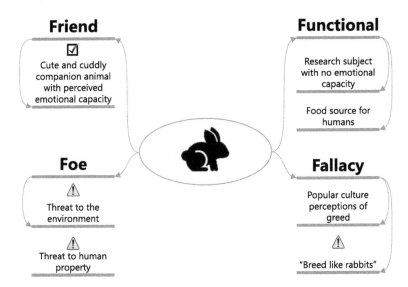

Figure 5.3 The Zoological Emotional Scale applied to the rabbit to identify where emotional capacity is attributed to this small mammal.

Scale to rabbits, it becomes evident that their welfare considerations are shaped by a convergence of competing perspectives around their perceived emotional capacity tempered by anthropomorphic social constructions and implications for more deserving creatures such as native wildlife.

THE DARK SIDE OF ANTHROPOMORPHISM

Anthropomorphism can provide an important entry point to an animal's mind, but there is also a darker side that has contributed to anthropomorphic selective breeding of infantile features in some animals. Zoologist, ethologist and ornithologist Konrad Lorenz identified that infantile (paedomorphic) physical features such as large foreheads and craniums, big eyes, bulging cheeks and soft contours prompted greater nurturing and attachment behaviours in humans, thus increasing the animal's survival chances (Hecht & Horowitz, 2015). Human attraction to infantile appearances has gained considerable research support over the years with the addition of facial flexibility as another appealing feature. Researchers Waller et al. (2013) reported that the more dogs in animal shelters were able to raise their inner brows which, in turn, enhanced the paedomorphic features of eye size and height, the faster they were rehomed.

According to Professor of animal ethics and welfare James Serpell (2002), who explored anthropomorphism and the "cute response" among companion animals, the effects on animal welfare can be problematic. Anthropomorphic selective breeding, defined as the selection of physical and behavioural traits that facilitate the attribution of human mental states to animals, has the potential to handicap animals physically and emotionally. Brachycephalic dogs, characterised by their large foreheads, large and low-lying eyes and bulging cheeks that fit Lorenz's baby schema perfectly, provide a popular example. Their infantile facial stimuli arouse positive emotions and nurturing responses in human adults, invoking and maintaining a strong attachment between human and dog, irrespective of the health impact on the dog.

Picture a French bulldog, round-faced and with large expressive eyes and an emotive flexible facial expression that calls out for love and attention. They also have a tendency to puff, pant, snort and snore due to their unnaturally shortened facial features, factors that are often overlooked as their snub-nosed appeal spreads through social media, memes, t-shirts and stationery. An inspection of Instagram's #frenchbulldog in 2023 reveals tens of millions of posts and images displaying these Brachycephalic dogs with their emotive expressions in human-like poses and activities. What these images fail to reveal is the high level of risk facing relatively young dogs as they cope with brachycephalic obstructive airway syndrome arising from soft tissues of the nose and throat being squeezed into unnaturally shortened upper jaws and noses. This can cause narrowing and resistance in the upper respiratory tract, prompting the World Small Animal Veterinary Association (2023) to describe the increasing number of short-nosed dogs as "an emerging canine welfare crisis" with some veterinarians expressing concern about the viability of continuing to breed dogs and cats with these infantile anthropomorphic features (Packer et al., 2019). It seems French bulldogs and other excessively short-nosed dogs including pugs, bulldogs, boxers, Cavalier King Charles Spaniels, Shih Tzus and mastiffs to name just a few can no longer be considered typical dogs, and yet their popularity continues despite potential health and welfare issues.

IMPLICATIONS

Anthropomorphism remains an important way to make sense of interactions with the animal world with a flow-on effect on conservation outreach and environmental education. The value of anthropomorphism requires a person to accept that they are merely an anthropologist in a foreign land who can only mix a pinch of what they would be feeling in that situation with a pinch of what they think it would be like for the animal in that situation to come up with a working hypothesis. This requires watching animals, accepting differences and resisting the temptation to fall back on familiar interpretations embedded in the human experience. However, it should not be perceived as a mechanism to privilege those animals deemed visually or emotionally worthy of human consideration at the expense of the small-eyed, slimy awkward creatures whose inner lives also matter.

The power of anthropomorphism to promote and sustain social practices, especially among young, developing minds, is important knowledge for parents, guardians and educators. Misrepresenting animal characters in media can give rise to unrealistic expectations that carry through to adulthood. When these expectations fail to eventuate, blame is often ascribed to the animal rather than the human environment, resulting in punitive measures to correct the animal's offending behaviour.

The Zoological Emotional Scale provides a lens through which to explore the role of anthropomorphism as one of the mechanisms fuelling socially learned perceptions of animal emotional capacity and the subsequent implications for the animal. Many fallacies are based on faulty anthropomorphism, sometimes dating back to childhood and never questioned in adulthood. Some ethical decision-making processes may thus be unconsciously based on learnings from animated movies, books and reinforced by social media. The Zoological Emotional Scale can provide one mechanism to explore the question—why do I think this way? Understanding the origin of attitudes and beliefs is a crucial part of the preparation phase of change.

The implications of anthropomorphic selective breeding cannot be under-estimated for the veterinary profession, animal rescue centres and human caregivers. Physical changes can impact the health and wellbeing of animals, as evidenced in Brachycephalic dogs and cats, resulting in unsustainable health expenses and potential abandonment. Of equal concern is the anthropomorphic selection of behaviours that can mutate the desired loyalty trait in dogs into a crippling dependency and hysterical separation anxiety when left alone.

REFERENCES

Bekoff, M. (1997). Deep ethology. In M. Tobias & K. Solisti (Eds.), *Intimate relationships, embracing the natural world.* Deep Ethology. https://web-archive.southampton.ac.uk/cogprints. org/161/1/199710001.html

Bekoff, M. (2007). *The emotional lives of animals: A leading scientist explores animal joy, sorrow, and empathy—and why they matter.* New World Library.

Bradbury, J.W. & Vehrencamp, S.L. (2011). *Principles of animal communication* (2nd ed.). Sinauer Associates.

Brrémaud, F. & Rigano, G. (2021). *Letters from animals.* Magnetic Press.

de Waal, F. (2001). *The ape and the sushi master.* Allen Lane - Penguin Books.

Epley, N., Waytz, A., Akalis, S. & Cacioppo, J.T. (2008). When we need a human: Motivational determinants of anthropomorphism. *Social Cognition, 26*(2), 143–155. https://doi.org/10.1521/ soco.2008.26.2.143

Garnett, A. & Kessing, K. (1994). *Easter Bilby.* The Australian Anti-Rabbit Research Foundation.

Grossblat, B. (2011). *Suicide Food.* http://suicidefood.blogspot.com/

Hebb, D.O. (1946). Emotion in man and animal: An analysis of the intuitive process of recognition. *Psychological Review, 53*, 88–106.

Hecht, J. & Horowitz, A. (2015). See dogs: Human preferences for dog physical attributes. *Anthrozoös, 28*(1), 53–163. https://doi.org/10.2752/089279315X14129350722217

Horowitz, A. (2010). *Inside of a dog: What dogs see, smell, and know.* Scribner Book Company.

Kulick, D. (2017). Human-animal communication. *Annual Review of Anthropology, 46*, 357–378. https://doi.org/10.1146/annurev-anthro-102116-041723.

Moe, A.M. (2016). The work of literature in a multispecies world. In S. Rice & A. G. Rud, (Eds.), *The educational significance of human and non-human animal interactions* (pp. 133–149). Palgrave Macmillan. https://doi.org/10.1057/9781137505255_9

National Film and Sound Archive of Australia (NFSA). 2017). *"Louie the Fly" Mortein Jingle (1962).* https://www.nfsa.gov.au/collection/curated/louie-fly-mortein-jingle-1962

Packer, R.M.A., O'Neill, D.G., Fletcher, F. & Farnworth, M.J. (2019). Great expectations, inconvenient truths, and the paradoxes of the dog owner relationship for owners of brachycephalic dogs. *PLoS ONE, 14*(7). https://doi.org/10.1371/journal.pone.0219918

Pandey, A. (2004). *Disney's designs: The semiotics of animal icons in animated movies. Sankota, 3*, 50–61.

Safina, C. (2015). *Beyond words: What animals think and feel.* Henry Holt and Company, LLC.

Safina, C. (2020). *Becoming wild. How animals learn to be animals.* Oneworld Publications.

Serpell, J.A. (2002). Anthropomorphism and anthropomorphic selection - Beyond the "cute response". *Society and Animals, 11*(1), 83–100.

Spada, E. (1997). Amorphism, mechanomorphism, and anthropomorphism. In R. Mitchell, Nicholas S. Thompson & H. L. Miles (Eds.), *Anthropomorphism, anecdotes, and animals* (pp. 37–49) (Part II). Suny Press.

The Secret Life of Pets. (2016). Directed by Chris Renaud. Universal Sony Pictures Home Entertainment Australia Pty Limited.

Waller, B.M., Peirce, K., Caeiro, C.C., Scheider, L., Burrows, A.M., McCune, S., et al. (2013). Paedomorphic facial expressions give dogs a selective advantage. *PLoS ONE, 8*(12). https://doi.org/10.1371/journal.pone.0082686

West, P. (2018). *Guide to introduced pest animals in Australia*. CSIRO Publishing.

World Small Animal Veterinary Association. (2023). *Brachycephalic Obstructive Airway Syndrome (BOAS)*. https://wsava.org/news/highlighted-news/brachycephalic-obstructive-airway-syndrome-boas/

Wynne, C.D.L. & Udell, M.A.R. (2013). *Animal cognition: Evolution, behavior and cognition* (2nd ed.). Palgrave UK.

6

GRIEF AND STRESS

In 1879, naturalist Arthur E. Brown commented on the unlikelihood of animals experiencing grief in *The American Naturalist* journal:

> Among the lower animals, with the exception of some domesticated varieties, any striking display of grief at the death or separation from an animal to the companionship of which they had been accustomed, has rarely been observed, and although a few statements of such occurrences have been made by different authorities, it is probable that the feeling of individual association, or friendship—if the term may be so used—partakes too much of an abstract nature to be sufficiently developed in them to retain much of a place in memory when the immediate association be once past.
>
> (p. 173).

Brown's next comments appeared almost contradictory as he described the behaviours of a chimpanzee male whose female mate had died, and whose "rage and grief" were painful to witness. The male chimpanzee attempted to rouse her, tearing his hair and emitting a yell of rage that transformed into a never-before heard "plaintive sound like a moan". He violently opposed attempts to remove her body from the cage. This was followed by a day of sitting still and quietly moaning.

Despite these behaviours, the naturalist concluded that "any high degree of permanence in grief of this nature belongs only to man" because only humans can counteract any prostrating effects by "an intelligent recognition of the desirability of repairing the injury suffered". Among lower animals the feeling would not develop because of its negative impact on survival, in particular, reproduction by impeding the need to look for a new mate or have more children.

The casual dismissal of true grief in any animal other than humans demonstrates an anthropocentric bias that, while gradually being eroded, still permeates the treatment of some animals in contemporary society. For a dairy farmer to acknowledge and respect that a cow may feel grief when forcibly separated from her newborn would destabilise the dairy industry where cows must bear calves in order to produce milk for human consumption. For a human to acknowledge that a companion animal would grieve if abandoned or relinquished to a shelter when they become old and inconvenient would cause unpleasant feelings of guilt. The contradictions within both the nineteenth-century article and the current treatment of some

DOI: 10.1201/9781003298489-7

animals depending on their socially constructed function reinforce the pressures to ignore or minimise animal grief.

This chapter focuses specifically on grief and loss and its relationship to perceived animal emotional capacity, leaving the concept of death and death awareness in animals to be explored in more depth in the next chapter. In this chapter, two important questions are probed: which animals are perceived as capable of experiencing grief for their kin, friends, freedom or lifestyle; and which animals are perceived as being worthy of human grief. The preceding chapters examined animal emotional capacity within the fundamental areas of attachment, family, reproduction and survival. In Chapter 4, we saw examples of empathy, altruism and direct reciprocity, inviting the logical inference that if an animal is prepared to go to great lengths to protect another, strong feelings must inevitably be triggered if the connection to that other is lost. Grief thus adds another potential dimension to the expressions of animal emotion.

WHAT IS GRIEF?

There is an over-abundance of definitions for human grief, most of which link grief to the concepts of loss and bereavement. The grief experience itself for human animals and nonhuman animals is a natural internalised response to loss, which can manifest cognitively, emotionally, spiritually and physically. Loss is the experience of losing something of personal relevance, whether that be through death or non-death related events, while bereavement is the period of sadness and loneliness after loss and resulting in the grief experience. Anthropologists studying human societies and cultures globally contend that most humans grieve loss to some degree and there is an almost universal attempt to regain the lost loved object (Worden, 2018). Part of human grief is acknowledging what used to be, feeling sadness at its loss and re-structuring life in its absence (Gold, 2020). This requires imagining a new future without the lost person or object. Most spiritual or religious traditions, cultures and societies provide parameters of grieving, with some even imposing a hierarchy of respectable and non-respectable grieving. This means grieving a companion animal or divorced partner may not rate as highly as grieving a child or parent. Human grief is not confined to death and can incorporate losses of future plans as well as anticipatory grief for a loss that is perceived as inevitable. However, when viewed through the lens of affective neuroscience, grief becomes a biological function that is no longer unique to humans.

Affective neuroscience positions grief as a primary emotion necessary for maintaining social bonds but which can, in some cases, cause significant impairment and anguish in an individual's life (Peña-Vargas et al., 2021). When Jaak Panksepp (1998) mapped out the foundations of human and animal emotion, he included Panic/Grief as the urge to be reunited with companions after separation, with a primary function of promoting social bonding. Focusing on mammals, Panksepp and Watt (2011) identified that oxytocin and prolactin (neurochemicals) are major social attachment and social bonding pathways in the mammalian brain. When these neuropeptides were present in high levels, they reduced feelings of separation distress. When an important attachment figure was lost, these neurochemicals were diminished leaving a feeling of withdrawal. Grief can therefore be viewed as a form of suffering, linked to the capacity to form and lose connections and relations. Grief, however, does not necessarily manifest in the same expressions and experiences across species, just as it does not present in the same way across humans and human cultures.

The tendency to transfer human models of grief and emotion to animals and use any identified behavioural differences as a rationale for claiming their absence in animal groups is flawed. To do so privileges human experience over animal experience and ignores the fact that humans remain animals themselves with different personalities and social support systems. Human grief counselling is premised on the acknowledgement of different approaches to grief at an individual and cultural level, making it logical that there will be different approaches to grief among animals.

Anthropologist Barbara King (2013) suggested an understanding of animal grief first required distinguishing it from other emotions. Grief could be suspected with the fulfilment of two prior conditions: firstly, there was evidence that the animals had chosen to spend time together beyond survival-oriented behaviours such as hunting/foraging or mating; and secondly, at the death of one animal, the other(s) changed their behavioural routine (for example, bodily posture, eating or sleeping habits, failure to thrive). The animal must also appear acutely distressed beyond sadness for a prolonged period, although how that acute distress manifested across species was not pre-determined. Unless a human had familiarity with an individual animal and could interpret changes in facial expressions, body language or other cues, the range of their reactions to loss could remain an ambiguous mystery, just as human expressions of grief can be unpredictable. Human reactions such as withdrawal, sadness, anger, physical violence or life-affirming behaviours demonstrate the complexity of this emotion in a species that can articulate their feelings to each other and reinforce the impossibility of discounting animal grief based on the lack of understanding.

CASE STUDY – EXPRESSIONS OF GRIEF

Like many people who live with, and love animals, I have grieved them and watched them grieve each other multiple times and in multiple ways. As neuroscientific evidence catches up with what people have known since taking animals into their hearts and their homes, my animal friends never cease to amaze me as they experience grief and sadness, and even anticipatory grief, in their own unique ways.

My animal family's most recent grief experience began with a thud when the elderly matriarch of my three cats failed to make the jump to my bed for her customary breakfast wake-up call. Indignity then concern flitted across her emotive face as 18-year-old Bella lay paralysed on the floor. Feline Aortic Thromboembolism, with the concerning acronym of FATE, had rendered Bella's rear legs useless, and it seemed inevitable that she would not recover from this sudden and serious event at her age. Her two housemates, Koko and Gypsy, began a process of what, in a human family setting, could be described as anticipatory grief—but for those sceptical of an animal's emotional capacity, could also be described as simply a reaction to change in the well-established household routine.

Twice Bella went to the veterinary clinic to be euthanised and twice her spirit and appetite for life rendered the veterinarian and me unwilling and unable to deprive her of another chance. Granting autonomy to the cat whose will to live remained strong despite the frailty of her body would eventually prove the right course of action.

As Bella remained immobilised and with the very real possibility of death, 17-year-old Gypsy, who had never missed a meal in her life, stopped eating. Gypsy's twice-daily hyperthyroidism medication, discretely hidden in the food she loved, remained untouched as she retreated to the furthest room from Bella to wait it out. Koko, the

Figure 6.1 Bella lies on a soft cushion, unable to move her back legs due to Feline Aortic Thromboembolism. Photograph supplied by the author.

carefree younger male, sat transfixed by the immobile Bella's side, only leaving her to
toilet and sing a low mournful song in the litter tray.

For five days, Bella lay on a soft bed on the floor, accepting assistance to eat and
toilet as she appeared to patiently await the return of movement in her hind legs. Bella's
stoic expression in Figure 6.1 was testament to her determination to live. On day six,
Bella staggered to the kitchen for the usual dinnertime routine. That same day Gypsy
emerged and demolished a full meal and Koko stopped singing to his litter tray. While
it is impossible to understand what Gypsy and Koko experienced emotionally, their
behavioural responses, each so different, tell a story of uncertainty and distress that in a
human we would not hesitate to call anticipatory grief.

There was no such reprieve when my young rescue dog Talyn lost his lifelong companion and role model, the 17-year-old blind and deaf Jesse. From a dog who could
hold his urine for hours, Talyn left puddles dotted around the house as he struggled to
come to terms with his solitary existence. After a month, the indoor puddles ceased,
until several years later when Talyn's human grandfather died and we had another few
weeks of urinary traps lurking around the house. Behavioural changes are common
outcomes of grief and if Talyn were a child wetting the bed, we would have no difficulty labelling this a grief response.

And finally, when Fatso, the jovial food-loving guinea pig succumbed to heart failure and was euthanised, his long-term friend, Max, screamed for two days. Guinea pigs
have a particularly piercing scream that sounds as if they are being eaten alive. To hear
this for two days was reminiscent of human wailing as a rite of mourning.

From a scientific perspective, the three examples above describe no more than a
correlational relationship between observed behaviours and an identifiable event. They
could also be dismissed as unvalidated anecdotes describing instinctive responses to
changes in routine or even anthropomorphised projections of human grief feelings

rather than the animals' emotional reactions to loss. Irrespective, the fact remains that these animals experienced something that changed their behaviour in a manner evocative of emotional pain and suffering and, therefore, should not be dismissed by scientific scepticism.

GRIEF EXPERIENCES

As humans gain greater access to animals' lives, be it first-hand through companion animals and domestic farmed animal sanctuaries or vicariously through social media and animal documentaries, anecdotal evidence of the meaningful bonds that animals form with each other and the impact when that bond is broken accumulates. Animals who develop lifelong bonds are inevitably at a greater risk of suffering at the death or removal of their companions or kin. Some risk their lives to be with the other individual, while others are perceived to have died of sorrow.

The neuroscientific basis of grief over the loss of a bonded partner has been examined within the context of prairie voles and attachment theory. Researchers Mary-Frances O'Connor and Saren Seeley (2022) describe how a part of the prairie vole's brain—the partner-approach neuronal ensemble in the nucleus accumbens—gets bigger after bond formation. Differences in the size of this ensemble predict attachment bond strength. This neural encoding is designed to trigger a physiological response to separation. It is specifically primed to produce the precursor of animal cortisol if one partner goes missing, motivating the vole to seek the partner and relieve stress through a reduction in cortisol and release of oxytocin and dopamine. When reunion is impossible—through death or captivity of the partner—the vole's cortisol levels remain high, reminiscent of the increased glucocorticoids associated with grief in humans and identified among baboons who have experienced loss (Engh et al., 2006). While this may seem to reduce grief to a biological level, it also validates the reality of what an animal may be feeling as equally meaningful to the biologically based human grief experience. There are many anecdotal examples of how animals demonstrate their negative feelings in culturally specific, nonhuman ways. Elephants provide one such unique example.

Reunion activities after separation among elephants include a display of tactile greetings and contact rubbing as a form of comfort (Rutherford & Murray, 2021). When an anticipated reunion does not happen, behavioural responses can be dramatic as the elephants' facial glands stream with emotion. Ecologist and author Carl Safina (2015) describes elephants as a "who animal" meaning that who dies and the survivor's relationship to the deceased is what triggers a response. The death of a matriarch can trigger strong psychological responses, as Safina described with the death of Eleanor.

When Eleanor, a matriarch living in Kenya's Samburu National Reserve, collapsed, another matriarch, Grace, lifted her back to her feet. Again Eleanor collapsed, and the distressed Grace, facial glands streaming with emotion, remained by her side until Eleanor died during the night. As other elephants—both family and friends—came to pay their respects over the next week, Grace remained by Eleanor's side. The strong need to be with a dead matriarch is not an isolated incident among elephants, but they have also been known to go on rampages destroying villages and lands in response to loss, reminiscent of human grief that can also manifest as uncontrollable anger. Maternal grief among elephants can have unexpected outcomes. In 2022, images of a mother elephant carrying around the body of her baby calf for days after it died featured in a news article (Thomson, 2022). The article included an implicit warning to resist the urge to anthropomorphise and imbue human-like emotions to this act, which could just as easily represent a lack of understanding that the calf was dead.

Human grief literature around the loss of an infant advises that every parent reacts differently to the death of their baby, and that for some, the opportunity to spend time with and create memories with the baby can be an important part of the grief journey. Grief is thus a personal experience, in which rules are developed along the way. It is feasible that for animals too, the behaviours associated with losing an infant are personal and open to multiple interpretations by human onlookers, none of whom can ever truly know the extent of what the animal is feeling.

Killer whales also share close and long-lasting family and maternal bonds. Breaking that bond had severe behavioural and emotional consequences for Tilikum who was described in Chapter 4, but also for the Pacific North West killer whale Tahlequah who was observed pushing her dead infant along the sea surface for 17 days in 2018 (Safina, 2020). A reported increase in killer whale newborn deaths potentially linked to increased toxic chemicals in the oceans and other human-induced global changes suggests that this behavioural display may not be an isolated incident in the future.

Among primates, stories of mothers carrying dead infants appear regularly in the literature. Primatologist Yukimaru Sugiyama reported approximately 2.3% of Japanese macaque mothers carried their dead infants for around a week. With a death rate of over 20% in the first year of life, conjecture arose as to whether there was an inability to recognise infant death, or equally plausible, an inability to accept infant death. Carl Safina (2020) posed a similar question about hormonally driven maternal urges preventing the recognition of death in Chimpanzee Ketie who carried her dead 2-year-old daughter for 10 days.

People have the power to capture and immortalise their loved ones through digital images and videos, something animals cannot achieve. When I know, or fear, one of my animals is dying, I capture their memory forever with many images on my camera phone. All my digital home screens are animals who have, or still live with me, and I know I am not alone in this tendency to maintain some form of continuing bond with a beloved animal. However, in some cultures, mourners who ponder too long and too intensely over images and videos risk diagnosis with Complicated Grief Syndrome. Grief specialist J. William Worden (2018) prefers the term complicated mourning, explaining that it is not the grief that causes problems as grief is a natural biological process, rather it is the mourning process that impedes the ability to move forward. While Tahlequah, Ketie and the Japanese macaques may be perceived as not recognising the death of their infants, they can also be re-positioned as grieving and not accepting the loss of their infant, conceptually no different to a human musing over images and objects from their dead loved ones for a socially unacceptable period of time. In the case of animals, they have no reminders other than the dead infant's body.

Behavioural changes indicative of separation loss were identified in a study of grief among a more familiar cohort, domestic dogs who had lived with another companion dog for many years (Uccheddu et al., 2022). While none showed the regression in toileting that my dog demonstrated at the loss of his 17-year-old mentor, the authors pointed out that even minor behavioural issues can become potential welfare problems as increased longevity among dogs increases the likelihood of them living for many years with a companion. The authors qualified their findings with the comment that they could not confirm these behavioural changes were grief, instead questioning the role of anthropomorphism in attributing specific feelings to the dogs' behaviours. Similarly, emotional contagion between the human caretaker and the remaining dog may also have contributed, with the dog mimicking the human's distress. It is noteworthy that the authors perceived anthropomorphism as detracting from the validity of their study, rather than providing a lens through which to interpret their findings.

Marc Bekoff expressed certainty that animals experience grief and cited many stories to demonstrate this in his influential book *The Emotional Lives of Animals* (2007). From the magpies paying respects to a dead friend by laying grass by the corpse, to the female fox who buried her dead mate, and to depressed wolves with heads and tails held low and mournfully crying, there are multiple examples of animals expressing psychological distress and emotional sadness at these losses. Importantly, Bekoff valued the role of anthropomorphism in interpreting these behaviours.

Irrespective of whether grief exists in animals or is an anthropomorphic projection, animal advocate and writer Teya Brooks Pribac (2021) drew attention to the fact that grief may be a luxury that not all animals can indulge in if they are to survive the loss of their child, their family or their habitat. In the wild, showing physical signs of grief over the loss of family may be perceived as a weakness and increase vulnerability to predators. The ability to grieve human encroachment or destruction of a habitat may be lost in the need to relocate and transition to a new habitat. More telling, and indicative of human dominance over animals, prolonged grief and stress in captive animals, including factory-farmed animals, may result in dispirited resignation and shutting down of the emotional and physical pain inherent in an inescapable situation. Inevitably, animals with whom humans identify will be more readily perceived as experiencing grief, and animals who do not exhibit human–centric grief behaviours will be more easily dismissed as not experiencing feelings of grief.

LOSS, STRESS AND POST-TRAUMATIC STRESS DISORDER

Feelings of grief experienced by animals and humans are not confined to losses associated with death. There are many other losses including loss of freedom, security, lifestyle, community or habitat that can leave an individual grieving, as well as at increased risk of a prolonged stress reaction. Fundamental to human dominance over animals is the capacity to deny their freedom, resulting in suffering and grief at an individual or group level. In a cruel twist of fate, in order to compensate for loss of a whole species due to human activities, a traditional response has been to lock samples of these animals in zoo environments. The resultant loss of freedom and inability to express normal behaviours with their family or social group can have long-term welfare implications. Prolonged captivity imposes suffering on sentient creatures, including persistent activation of the hypothalamic–pituitary–adrenal axis and behaviours akin to the human grief response and indicative of psychological trauma (Pierce & Bekoff, 2018). One physical manifestation of activation of the hypothalamic–pituitary–adrenal axis can be stereotypies, the stereotypic behaviours that indicate all is not well in the animal's central nervous system and that pathological changes to the brain are underway. These actions can be rocking, pacing back and forth, pulling out hair or feathers, scratching or biting themselves, or regurgitation and reingestion. Higher levels of faecal glucocorticoid concentrations—an indicator of increased stress in animals and humans alike—have been identified in animals displaying stereotypies. Certain animals who require freedom to roam through specific environments do not do well in captivity and confinement, including elephants, bears, wolves, whales, dolphins, chimpanzees, orangutans, lions and tigers to name a few. While the solitary captive elephant in Figure 6.2 has food and safety, he is lacking family life and the freedom to roam widely, leaving him at risk of developing a stress–related grief response long-term.

CASE STUDY – HAPPY THE ELEPHANT

Since the 1970s, the ironically named Asian elephant Happy lived in isolation at the Bronx Zoo (New York, US). Moving between an outdoor lot barely more than an acre (less than half a hectare) and a barred cage indoors, Happy's relentless stereotypic behaviours of trunk swinging, swaying and lifting her feet indicated the chronic stress of not having her physical or psychological needs acknowledged. Stress associated with grief can be difficult to separate from everyday stress, supporting the notion that the stereotypies frequently exhibited by captive elephants may be a combination of both frustration at confinement, together with grief over loss of freedom, family and opportunity to live a free elephant's life. Happy's restless stereotypies, so common among captive elephants, seemed to clearly demonstrate her long-term unhappiness.

In 2022, the Nonhuman Rights Project filed a writ of *habeas corpus* with the state of New York seeking the court's recognition of Happy's right to be released from confinement and relocated to an elephant sanctuary, thus ending 16 years of isolation (Comstock et al., 2022). The 2022 writ of *habeas corpus* was unsuccessful as Happy was deemed not to be a person. That meant her detainment in the Bronx Zoo was not illegal, and she could legally remain there. This important case uncovered human fear in the commentary supporting the ruling, with statements citing the enormous destabilising impact on modern society and the effect this would have on the way humans

Figure 6.2 An elephant in captivity stands alone in his enclosure. Photograph supplied by the author.

interact with animals if they were given personhood. It is likely that this fear was also underpinned by the way humans treat animals in general and the far-reaching implications of having to re-imagine social norms should all animals be perceived as sentient beings with personhood.

Zoos may also promote grief behaviours in animals when moving them around for breeding purposes, as part of an exhibit or to balance numbers. Irrespective of the reasons, if an animal has formed a social attachment, moving them from familiar surroundings with familiar cohorts can be distressing and prompt a grief experience, not unlike the child whose parents constantly relocate for work reasons.

WHEN ANIMALS GRIEVE THEIR HUMANS

There are many anecdotal stories of companion animals grieving their human carers, some even immortalised in statues such as Greyfriars Bobby. Bobby's headstone reads: "Greyfriars Bobby – died 14th January 1872 – aged 16 years – Let his loyalty and devotion be a lesson to us all", although the facts behind the story vary somewhat. Hachikō the Akito has a similar back story, with a statue erected at Shibuya train station for the dog who waited there between 1925 and 1935 for the professor who died in his office and never came home. Following Hachikō's death, he was cremated and his ashes were placed next to his professor's grave.

More obscure and personal stories are multiple, with statistics from one animal rescue service estimating that thousands of animals come into their care following the death or relocation to aged care of their human. Often old themselves and abruptly torn from the only carer they have ever known, these animals demonstrate confused and distressed behaviours such as aggression, refusal to eat, toileting issues, lack of trust and withdrawal, all of which put them at risk of non-adoption and euthanasia.

Sadness and disruption at the loss of a human may not be limited to domestic animals. The story of elephants mourning the death of the "elephant whisperer" Lawrence Anthony is legendary, with two herds of elephants who had not visited the family compound for months reported to have travelled half a day to arrive shortly after Anthony's death. They loitered for 2 days and then returned to the bush, suggesting that these creatures, known to mourn their own dead, were feeling emotions for the death of the man who had rescued and rehabilitated them.

ANIMALS WORTHY TO BE GRIEVED

There is little doubt that companion animals are firmly located in the domain of animals worthy of human grief. Considerable research supports the notion that grieving the loss of a companion animal is a traumatic experience for some people (for example, see the systematic review by Kemp et al. 2016). Much of this research suggests that where an attachment relationship has formed, human grief for an animal can be as severe as for another human. However, social norms do not always position grief for an animal as of equivalence with a human loss, leaving the bereft person forced to comply with social expectations and grieve alone, or turn to those with whom they feel safe to share their sorrow.

People are necessarily selective in their grief for animals. Mourning the loss of a companion animal whom they perceive as friend or kin is qualitatively different to feelings about the death of other members of their companion animal's species, or other species with whom they share a home. Rodents, reptiles and insects exist in a subliminal space where their removal or invisible death by poison is actively sought without remorse or feelings of grief. It can also be hard to grieve an animal where there is uncertainty around its capacity to suffer, or there is a lack of familiarity. Sometimes lack of familiarity can be overcome by an individual animal whose public life and death experiences provide a catalyst for human grief extending to a species.

In 1993, killer whale Keiko shot to stardom in the feel-good movie *Free Willy*. Ten years later he was dead, after apparently failing to adapt to the new life in the wild that the movie had inspired. Perceived as an emotional and loving creature following the movie, the publicity around Keiko's death brought on an outpouring of grief, anger and accusations. Globally, beached whales attract media attention as volunteers and professionals alike seek to refloat these sentient creatures. People will spend hours and days with these whales, sending images around the world as they grieve the ones who succumb to the trauma. Research into these aquatic mammals has seen them re-positioned from an aquatic animal slaughtered for body parts to an intelligent, sentient warm-blooded creature worthy of human grief and protection. When Japan re-commenced whaling in 2019 with strict quotas, the world responded angrily to footage of a Minke whale forcibly held under water for 20 minutes to drown, then sold for meat (Drury, 2021).

Social media and global connectivity have done much to increase the visibility of distant animals, prompting outpourings of communal grief for animals never met. In July 2022, the death from old age of Tricia, a 65-year-old elephant who had lived at the Perth Zoo in Australia since 1981 captured Australia's attention. City of Perth Council House lit up with an image of an elephant and a memorial walk for the public was established with pictures telling the story of Tricia's life and the opportunity to sign one of Tricia's trunk "kiss" artworks and leave messages of condolence for the memorial book (Richards, 2022).

On a wider scale, Cecil the lion's death by a recreational big-game hunter prompted grief and rage at a global level. Cecil, an African lion from the Hwange backwaters of Zimbabwe, was cast out from his pride after an epic battle that saw the death of his brother. Cecil went on to become a great favourite among Wilderness guides and an Instagram celebrity by 2009. Siring 18 cubs, Cecil stayed in power for several years, moving between the protected national park and occasionally straying into the forests that formed part of a trophy-hunting concession. By 2013, Cecil had been exiled again, then returned triumphant after teaming up with a former enemy, Jericho. Cecil's reputation grew as he became one of "the first lion superstars of the digital media age" (Ham, 2020, p. 97). In 2014, Cecil was illegally shot with an arrow, and eventually tracked and killed the next morning. The social media world erupted in a show of anger and grief, supported by celebrities who tweeted their support of Cecil to millions of followers, while Jericho, Cecil's unlikely partner, roamed the area for days, roaring and calling for his friend. The suffering of individual animals—especially if they have a name and a story—has the potential to promote more human grief than mass killings. While the death of Cecil raised awareness of hunting and the ongoing need for conservation, the larger issue of loss of habitat through human expansion that indirectly provided the circumstances for Cecil's death remained largely unspoken at this time.

Similarly, the suffering and death of farmed animals, both vertebrate and invertebrate, remains hidden from the consumer until individual animals are given a name and a face. Hundreds of thousands of nameless sheep are transported long distances in Australia to reach saleyards, abattoirs and export ports. Travel times can be long and stressful in a continent spanning 7.6 million km², causing suffering and deaths along the way as animals are separated from their familiar environments, handled by strangers and experience overcrowding, water and food deprivation and exposure to extremes in temperature. The Australian *Animal Welfare Standards and Guidelines – Land Transport of Livestock* (Agriculture Victoria, 2020) was developed to ensure basic animal welfare requirements of animals during transport, but this neither acknowledges nor alleviates the individual confusion and suffering of nameless sheep, cows, pigs or chickens crowded together on the trucks.

But what a difference a name and some caring humans can make. Farmed animal sanctuaries rescue one animal at a time and bestow on them a name and an identity. They are transformed into unique individuals with a life story and the right to be grieved when they die. In 2015, a little lamb born with crippled front legs arrived at the Australian farmed animal sanctuary, Edgar's Mission. Named Lemonade, and allotted the catchy phrase, "When life gives you lemons—make Lemonade", the tiny lamb's long rehabilitation that involved physiotherapy, splints, prosthetics and good nutrition went global via social media. I was privileged to become Lemonade's best buddy through the sanctuary's sponsorship programme and rejoiced in his transition from the tiny bow-legged lamb in a multi-coloured jacket in Figure 6.3 to a larger-than-life personality who was a friend to humans and animals alike at the sanctuary.

In October 2022, I received an unwelcome email from Edgar's Mission—Lemonade had died. I cried as I took in the meaning behind the words:

> [….] the sad news that his life on this earth has come to an end. A somewhat hasty departure that appeared to come from nowhere, as our hearts and veterinary science are still determining the cause […] Oh dearest Lemonade, through tears of heartache we type, but we know they shall pass. However, your memory never will. We loved you then and shall love you forever more.
>
> (Personal email correspondence from Edgar's Mission)

As an individual with a name, Lemonade earned the right to be grieved by humans who had never met him, in sharp contrast to the hundreds of thousands of sheep who die in factory-farm conditions, during road or sea transport, or simply because they are a commodity whose usefulness runs out when they are born crippled or become sick. When viewed within the Zoological Emotional Scale, Lemonade achieved friend status and his emotional capacity to enjoy life to the fullest was never in doubt. To contemplate the suffering of hundreds of thousands of sheep not so fortunate as Lemonade can become overwhelming for some people who may feel powerless to change the status quo of these functional animals.

In March 2021, a stranded cargo vessel in the Suez Canal led to what has been described as the "worst maritime animal welfare tragedy in history" with an estimated 200,000 live animals stuck on ships with insufficient food and water (Gherasim, 2021). Many were sheep travelling from Romania to the Persian Gulf for slaughter. Knowledge of the suffering and lingering death of so many sentient creatures should have been overwhelming if equated with the feelings evoked by Lemonade's death. However, the phenomenon of psychic numbing goes some way to explain the different

Figure 6.3 The young Lemonade in a warm jacket after first arriving at Edgar's Mission. Photo provided by Edgar's Mission Farmed Animal Sanctuary and reproduced with their permission.

human grief responses to animal death and suffering on this massive scale. Serving as a flawed protective mechanism, psychic numbing instils a sense of indifference and removal of personal responsibility when a disaster becomes too large to contemplate or to remedy. Mother Teresa (1910–1997), the Catholic nun who devoted her life to serving the poor and destitute, commented: "If I look at the mass, I will never act. If I look at the one, I will" (Butts et al., 2019). Lemonade and his friends at Edgar's Mission and other sanctuaries around the world draw attention to the one and the possibility of enacting change for one animal at a time.

INVERTEBRATE ANIMALS

When considering invertebrates, it would be presumptive to dismiss the possibility that they feel anything akin to grief on the basis that expressions of grief have not been identified in this group, nor has attachment as applied to mammals been identified in invertebrates. Some invertebrates are very connected to the collective, with communal insects such as honeybees presenting as a whole superorganism, comprising many worker bees performing bodily functions, an organ responsible for conception and birth (the queen) and another organ for fertilisation (drones). The collective takes on the thinking aspect and all parts communicate and know their roles, much like parts of the mammal body. Damage to any part of the collective can have implications to the organism, as evidenced by the Varroa destructor external parasitic mite. This mite has wrought havoc on honeybee colonies by feeding and reproducing on larvae and pupae in the developing brood, resulting in malformation and weakening of the bee and transmission of viruses (Imhoof & Lieckfeld, 2014).

Wasps and termites similarly form colonies, but it is humans who seek to actively destroy them when they are perceived to be encroaching on human territories. Despite the diversity of invertebrates, they are often excluded from welfare considerations unless their destruction has a negative impact on humans (such as the commercial effects of bee loss). Their value, and therefore their worthiness to merit human grief, is assessed instrumentally in terms of the broader ecological or economic services they provide rather than ethical consideration of their individual right to life. Other factors that come into play when assessing their worthiness to live and be grieved include a lingering perception that invertebrates are lower in the hierarchy of living creatures; human biases that skew mental state attributions to unfamiliar, disgust-provoking creatures and therefore their worthiness to be grieved; and the fact that there are just so many of them (Mikhalevich & Powell, 2020).

The need to not only grieve insects, but also take action to prevent their loss, may be more important than first thought in the face of mounting evidence of the accelerated disappearance of these tiny essential creatures. In 2017, a research article authored by a group of European scientists made the extraordinary claim that flying insect biomass had dropped by 76% in just over 27 years in some protected nature areas in Germany (Hallmann et al., 2017). If this was true, the flow-on effects on pollination, nutrient cycling, food sources for birds, mammals and amphibians, and healthy ecosystem functioning in general would be devastating. Inevitably, the validity of these claims and the scale of insect decline were disputed, often based on the dearth of long-term rigorous research. Anecdotal evidence, however, continued to filter through.

During my childhood, patches of shimmering brown, yellow and pink would dot the ground in the lead up to the summer holiday season. All along the east coast of Australia, the colourful Christmas Beetles, members of the scarab family, emerged in November and December to herald in the festive season with their iridescent exoskeletons. Stories from the 1920s and 1930s describe tree branches bending under the weight of these massed beetles and the noise of their whirring wings in confined spaces like the sound of distant aircraft (Millman, 2022). Fast forward to the twenty-first century, and the Christmas beetle seems to have abandoned the festive season and the country. The encroachment of brick and concrete residential expanses on their former habitats of leafy green eucalypt trees cannot sustain this dwindling species, and so once a year those Australians who remember these glowing insects mourn their disappearance. In these same areas, other insects such as mosquitoes continue to proliferate, suggesting that the loss of biodiversity among insects is selective and some insects will always survive. Human interference in the environment cascades in unanticipated directions.

The loss of biodiversity among mammals, birds, fish and reptiles is routinely documented, grieved and extraordinary measures taken to restore what was lost. In 2018, the death of Sudan, the world's last male northern white rhinoceros, was announced with great sadness. Leaving behind two female white rhinos, the species' last hope lay in combining eggs from the two females with stored northern white rhino semen and implanting into surrogate southern white rhino females through in vitro fertilisation (IVF) techniques (Gichohi, 2019). Sudan was cited as a key moment for conservationists globally as he became a symbol of grief and loss for a changing world. The loss of insects has the potential to negatively impact humanity in ways far greater than Sudan's death, providing yet another reminder of the need to not only grieve biodiversity loss but also to use these feelings as a catalyst for change.

IMPLICATIONS

Acknowledging that some animals have the capacity to grieve and be grieved has far-reaching welfare implications for animals and humans. From nineteenth-century naturalist Arthur E. Brown's cavalier dismissal of the chimpanzee's grief that he had just described, to the ruling against Happy the elephant's personhood to avoid the destabilising effect on modern society, it becomes clear that humans have much to lose if animals gain the right to grieve their kin, their friends and their freedom. Moving away from a human-centric perception of acceptable grief behaviours and embracing interspecies equality poses two scenarios for humans. Being the direct or indirect cause of grief in an animal but unwilling or unable to alleviate this can result in cognitive dissonance and the need to find strategies to reconcile these unpleasant feelings. Alternatively, where empathic humans take on board the pain and suffering of animals, often brought about by human actions, the ongoing extended experiences of grief place them at risk of vicarious trauma and compassion fatigue.

Ironically, both these implications relate to the effect on humans, rather than on the animals themselves, and both feelings can be alleviated by a change in treatment of animals. To avoid feelings of cognitive dissonance or vicarious trauma and compassion fatigue, the solution lies in the behaviours of the humans who refuse to acknowledge and validate the existence and worthiness of other-species grief. As the conversation moves from "Do animals grieve?" to "How do animals grieve?" (Pierce, 2018), the next logical question is, what can humans do to acknowledge and support animal grief?

The first step lies in watching animals and accepting differences. In traditional human grief research, there has been a logical positivism tendency that has drawn from the scientific paradigms of ethology, medicine, psychiatry and psychology. Increasingly, this is being replaced by a move towards a post-modern, relativist emphasis on the uniqueness of each person's grief, succinctly summed up in the phrase "there is no right or wrong way to grieve" (Wilson et al., 2014, p. 36). Post-modern researchers still acknowledge a need for research into commonly reoccurring patterns to supplement the unique in evidence-based grief counselling practice, but the mercurial nature of grief becomes increasingly evident when considering that a grieving person may feel sad, angry, anxious, shocked, regretful, relieved, overwhelmed, isolated and numb, sometimes all on the same day. If there is so much complexity within a familiar species, how then can humans presume to understand this process within other species?

Humans working in rescue work deal with animal suffering and grief on a daily basis and understand the patience that is required to allow that grief to run its course. Time may not be on the animal's side, as behavioural manifestations of grief can result in the animal being euthanised before the symptoms subside. Acceptance of difference, time and patience sound like an easy solution to accepting and responding to animal grief, but this may not be an option when limited resources and competing human needs are given precedence over animal emotions. A paradigm shift would see observation being accepted as a valid starting point for the acceptance of grief in a range of other species.

REFERENCES

Agriculture Victoria (2020). Animal Welfare Standards and Guidelines – Land Transport of Livestock. https://agriculture.vic.gov.au/livestock-and-animals/animal-welfare-victoria/livestock-management-and-welfare/land-transport-of-livestock-standards-and-guidelines

Bekoff, M. (2007). *The emotional lives of animals: A leading scientist explores animal joy, sorrow, and empathy—and why they matter.* New World Library.

Brooks Pribac, T. (2021). *Enter the animal: Cross-species perspectives on grief and spirituality.* Sydney University Press.

Brown, A.E. (1879). Grief in the Chimpanzee. *The American Naturalist, 13*(3), 173–175. https://www.jstor.org/stable/2448772

Butts, M.M., Lunt, D.C., Freling, T.L. & Gabriel, A.S. (2019. Helping one or helping many? A theoretical integration and meta-analytic review of the compassion fade literature. *Organizational Behavior and Human Decision Processes, 151*, 16–33. https://doi.org/10.1016/j.obhdp.2018.12.006

Comstock, G., Lerner, A. & Singer P. (2022, May 17). Why the court should free happy. Inside Sources. https://insidesources.com/why-the-court-should-free-happy/

Drury, F. (2021, January 31). Japan whale hunting: 'By-catch' rule highlighted after minke death. *BBC News.* https://www.bbc.com/news/world-asia-55714815

Engh, A.L. Beehner, J.C., Bergman, T.J., Whitten, P.L., Hoffmeier, R.R., Seyfarth, R.M. & Cheney, D.L. (2006). Behavioural and hormonal responses to predation in female chacma baboons (*Papio hamadryas ursinus*). *Proceedings of the Royal Society B: Biological Sciences, 273*(1587), 707–712. https://doi.org/10.1098/rspb.2005.3378

Gherasim, C. (2021, March 30). Some 200,000 animals trapped in Suez canal likely to die. *EUObserver.* https://euobserver.com/world/151394

Gichohi, N. (2019, March 19). *Remembering Sudan: What the loss of the last male northern white rhino means for all of Africa's rhinos.* African Wildlife Foundation. https://www.awf.org/blog/remembering-sudan-what-loss-last-male-northern-white-rhino-means-all-africas-rhinos

Gold, J.M. (2020). Generating a vocabulary of mourning: Supporting families through the process of grief. *The Family Journal: Counseling and Therapy for Couples and Families, 28*(3), 236–240. https://doi.org/10.1177/1066480720929693

Hallmann, C.A., Sorg, M., Jongejans, E., Siepel, H., Hofland, N., Schwan, H., Stenmans, W., Müller, A., Sumser, H., Hörren, T., Goulson, D. & de Kroon, H. (2017) More than 75 percent decline over 27 years in total flying insect biomass in protected areas. *PLoS ONE, 12*(10), e0185809. https://doi.org/10.1371/journal.pone.0185809

Ham, A. (2020). *The last lions of Africa.* Allen & Unwin Publishers.

Imhoof, M. & Lieckfeld, C-P. (2014). *More than honey. The survival of bees and the future of our world.* Greystone Books Ltd.

Kemp, H. R., Jacobs, N. & Stewart, S. (2016). The lived experience of companion animal loss: A systematic review of qualitative studies. *Anthrozoos, 29*(4), 533–557. https://doi.org/10.1080/08927936.2016.1228772

King, B.J. (2013). *How animals grieve*. University of Chicago Press.

Mikhalevich, I. & Powell, R. (2020). Minds without spines: Evolutionarily inclusive animal ethics. *Animal Sentience, 29*(1). https://doi.org/10.51291/2377-7478.1527

Millman, O. (2022). *The insect crisis: The fall of the tiny empires that run the world*. W.W. Norton & Company, Inc.

Nonhuman Rights Project. (2023). *Client, Happy (Elephant)*. https://www.nonhumanrights.org/client-happy/

O'Connor, M. & Seeley, S.H. (2022). Grieving as a form of learning: Insights from neuroscience applied to grief and loss. *Current Opinion in Psychology, 43*, 317–322. https://doi.org/10.1016/j.copsyc.2021.08.019

Panksepp, J. (1998). *Affective neuroscience: The foundations of human and animal emotions*. Oxford University Press.

Panksepp, J. & Watt, D. (2011). What is basic about basic emotions? Lasting lessons from affective neuroscience. *Emotion Review, 3*(4), 1–10. https://doi.org/10.1177/1754073911410741

Peña-Vargas, C.I., Armaiz-Pena, G.N. & Castro-Figueroa, E.M. (2021). A biopsychosocial approach to grief, depression, and the role of emotional regulation. *Behavioral Sciences, 11*(8), 110. https://doi.org/10.3390/bs11080110

Pierce, J. (2018, August 24). Do animals experience grief? *Smithsonian Magazine*. https://www.smithsonianmag.com/science-nature/do-animals-experience-grief-180970124/#:~:text=A%20growing%20body%20of%20scientific,for%20or%20ritualize%20their%20dead

Pierce, J. & Bekoff, M. (2018). A postzoo future: Why welfare fails animals in zoos. *Journal of Applied Animal Welfare Science, 21*(suppl1), 43–48. https://doi.org/10.1080/10888705.2018.1513838

Richards, N. (2022, July 9). Perth Zoo elephant Tricia and how you can pay your respects on a memorial walk. *The Western Australian*. https://www.perthnow.com.au/local-news/perthnow-central/perth-zoo-elephant-tricia-and-how-you-can-pay-your-respects-on-a-memorial-walk-c-7463739

Rutherford, L. & Murray, L.E. (2021). Personality and behavioral changes in Asian elephants (*Elephas maximus*) following the death of herd members. *Integrative Zoology, 16*, 170–188. https://doi.org/10.1111/1749-4877.12476

Safina, C. (2015). *Beyond words: What animals think and feel*. Henry Holt and Company, LLC.

Safina, C. (2020). *Becoming wild. How animals learn to be animals*. Oneworld Publications.

Sugiyama, Y., Kurita, H., Matsui, T., Kimoto, S. & Shimomura, T. (2009). Carrying of dead infants by Japanese Macaque (Macaca Fuscata) mothers. *Anthropological Science, 117*, 113–119.

Thomson, J. (2022, November 3). Heart-wrenching photos show elephant mom carries body of dead calf for days. *Newsweek*. https://www.msn.com/en-au/news/world/heart-wrenching-photos-show-elephant-mom-carries-body-of-dead-calf-for-days/ar-AA13GNpR?ocid=winp1taskbar&cvid=c0b3d9bedc4a4810bdedbec8d4cf9860

Uccheddu, S., Ronconi, L., Albertini, M., Coren, S., Pereira, G.D.G., de Cataldo, L., Haverbeke, A., Mills, D.S., Pierantoni, L., Riemer, S., Testoni, I. & Pirrone, F. (2022). Domestic dogs (*Canis familiaris*) grieve over the loss of a conspecific. *Scientific Reports, 12*, 1920. https://doi.org/10.1038/s41598-022-05669-y

Wilson, J., Gabriel, L. & James, H. (2014). Observing a client's grieving process: Bringing logical positivism into qualitative grief counselling research. *British Journal of Guidance & Counselling, 42*(5), 568–583. https://doi.org/10.1080/03069885.2014.936823

Worden, J. W. (2018). *Grief counseling and grief therapy: A handbook for the mental health practitioner* (5th ed.). Springer Publishing Company. https://doi.org/10.1891/9780826134752

7

AWARENESS OF DEATH

For all animals, and humans are animals, life and death are sequentially connected. Death after giving life may be immediate, such as the drone bee whose sex organs split open and are torn from his body immediately after providing the queen bee with millions of sperm cells to store and hatch as required. For others, such as the octopus, it comes when offspring become independent and no longer need their mother. A small number of humans can live 100 years or more, while Jonathan, a giant Seychelles Tortoise on an island in the South Atlantic Ocean, entered the Guinness Book of Records in 2022 at 190 years old (Atwal, 2022).

Despite the tendency for humans to believe they hold a monopoly on understanding the inevitability and meaning of death, there is growing evidence that some animals do in fact have an awareness and understanding of death. In his ground-breaking book *The Question of Animal Awareness* (1976), professor of zoology Donald Griffin questioned the absolute belief by the likes of Miller et al. (1960), Langer (1962) and others that humans were the only animals that could be aware of their own future death. He gently refuted this absolutism by drawing attention to some key anomalies:

> But I suggest we pause and ask just how anyone knows this. What sort of evidence is available either pro or con? Suggestive inferences can be based on the clear demonstration that many social animals recognise each other as individuals, and on the observation that some animal mothers show signs of distress over the corpses of their dead infants [...] How can we judge whether an animal may experience any notion of its own death after observing the death of companions (Cowgill, 1972)? The available, negative evidence supports at most an agnostic position.
>
> (1981, p. 104)

Humans control life and death for many animals. This can be a direct connection, including animals bred to be slaughtered for human consumption; or animals whose sole function is to participate in research studies where the endpoint is humane killing; or the mass destruction of animals labelled as pests. It can also be the decision to euthanise an animal, for medical or convenience reasons; or it may be through hunting, fishing or an ongoing and long-term degradation of the environment and climate, making it impossible for native animals to survive. Lost and abandoned fishing gear, deadly to marine life, provide one such example as they pollute ocean waters and

DOI: 10.1201/9781003298489-8

inflict death on countless future generations. Dumped nets, lines, pots and traps, the so-called "ghost gear", trap and entangle small fish, crustaceans, sea turtles, whales and sea birds. This is in addition to the estimated 8 million tonnes of plastic waste, broken down into microplastics, entering the oceans annually (Uy & Johnson, 2022). In fact, killing animals has become a multi-million dollar industry, spawning pest control companies and chemicals of mass destruction, especially for invertebrates.

Humans not only control the time of death, they have the power to determine whether it is a good death or bad death depending on perceptions around that animal's worthiness and emotional capacity. The cat, who occupies differing positions on the Zoological Emotional Scale, may be euthanised in the arms of their human with all the rituals and rites associated with kinship. In a different geographic location, the cat may be mass slaughtered as a food source, abandoned to live or die when no longer required as an item of property, or culled by chemicals when deemed a threat to native animals. Human subjective perceptions of an animal's capacity to feel emotions may thus play a key role in their death, not just their life. With growing awareness and acceptance of consciousness, sentience and emotional capacity in vertebrate and invertebrate animals, awareness of death is an integral part of examining animal emotions.

In the preceding chapter, the concept of grief was introduced within the context of which nonhuman animals are worthy of human grief, and which animals are perceived as being capable of experiencing grief. Grief and loss in that chapter were not confined to death alone. Drawing on the assumption that some animals do experience grief although not necessarily in a human-centric way, this chapter examines what death means to the animal, and the different ways that humans, as the more dominant species, enact death on animals.

WHAT DOES DEATH MEAN?

Many humans struggle with the concept of death resulting in death anxiety, a term used to describe the fear of death or dying that can influence a person's experience, feelings and conduct (Yalom, 1980). Cultural anthropologist Ernest Becker's Terror Management Theory proposed that a defining characteristic of human beings was their self-awareness about being alive which could lead to wonder and joy. However, this feeling was tempered by overwhelming dread at the realisation that life was accompanied by inevitable and largely uncontrollable death (Greenberg, 2012). Existential philosophers positioned death anxiety as a profound human apprehension to be strongly defended against and denied and consuming considerable life energy while doing so. Death can remain largely in the background of many people's lives, but it remains ready to be brought to the forefront depending on circumstances. According to psychiatrist, psychotherapist and psychoanalysis Robert Langs (2004), death anxiety is therefore "the most powerful unconscious psychodynamic dynamism in present-day emotional life" (p. 32). With the emphasis on humanity's psychological relationship to death, this serves as yet another concept to maintain human exceptionalism with the inference that animals lack sufficient self-awareness to achieve this psychic relationship with death.

All these theories around death include both a cognitive and emotional component, with the cognitive striving to subdue the more instinctive emotional aspect that manifests as anxiety and mental disturbance if released. Uncertainty around an animal's capacity for cognitive appraisal of death, including awareness of their own mortality, traditionally excludes animals from any existential discussion of death, banishing their life and death experiences to a more basic fight or flight response.

The stereotypical actions of eusocial insects triggered by certain chemical characteristics of corpses also support this basic response. Ants recognise death when they detect the smell of triglycerides on their dead conspecifics and carry them from the nest to an "ant cemetery" to keep the nest hygienic (Monsó & Osuna-Mascaró, 2021). Coating a live ant in the death-chemical Oleic Acid can result in the ant being carried off to the cemetery, legs waving wildly in a show of life. Burying behaviour among rats is similarly triggered after smelling putrescine or cadaverine, two foul-smelling compounds given off by the decaying rat carcass. An anaesthetised conspecific or pieces of wood will also be buried if sprinkled with these compounds (Monsó, 2022).

Among solitary animals such as octopuses, death is intimately connected to reproduction and arrives after offspring hatch. After laying thousands of eggs, the female will stop eating in favour of tending to the eggs, chasing away predators and fanning the developing offspring with oxygen-rich water. As the offspring leave their egg cases, the exhausted mother commences her death throes, completely ignored by her offspring, and often to be snatched up by a passing predator. In 1977, Jerome Wodinsky found that surgically removing optic glands from female Caribbean two-spot octopuses reversed this behaviour with the female abandoning her babies and choosing life over death, although ongoing research suggests the connection is more complex than just removing glands (Yan Wang & Ragsdale, 2018).

In all these examples, death is presented as a biological function with no emotional activation involved. However, there are also several mammal and avian species that show distinctive and more flexible responses to death. Grief over death of familiars was examined in Chapter 6 with numerous observational cases of animals showing changed behaviours during the pre- or post-death period, suggesting that they have an understanding that a member of their group will leave or has left permanently. Most of these responses required there to be a corpse that they could see, touch and enact rituals such as covering the corpse with grass, twigs, branches or dirt.

Humans may find it easier to relate to death awareness among primates, making Carl Safina's (2020) description of the death of an adult chimpanzee heart-breaking, but not unexpected. As the female lay dying, she was surrounded by two dozen chimpanzees making strange calls. The alpha male kept his distance, demonstrating signs of fearfulness. Finally, they all left, leaving only the chimpanzee's two children uttering peculiar sounding screams leading up to their mother's death. Marc Bekoff's (2007) description of the death ritual of a female fox who buried the corpse of a male fox may be more open to conjecture for those who choose to be sceptical. Bekoff himself sought advice on this phenomenon, as he describes: "A few hours later I went to see the carcass, and it was totally buried. No one to whom I have spoken, naturalists or professional biologists, has ever seen a red fox bury another red fox" (p. 64).

Reading Bekoff's words was reminiscent of a ritual enacted by my dog and a dead kitten many decades ago. During the years my father provided sanctuary to the university cats, our dog Johnny Belinda (JB) acted as a surrogate father to a number of kittens born on the back verandah. He became particularly enamoured with a group of five kittens born to a young mother barely out of kittenhood herself. The mother's youth and food shortages during her pregnancy meant the kittens started life at a deficit, seeming to trigger an even stronger nurturing instinct in JB. The death of one kitten saw JB lift him from the mother's bed and carry him around the back garden until seeming to settle on a suitable location. Gently lowering the kitten's body, he dug a hole, buried the corpse and then remained lying on the nearby grass for several minutes in a watchful vigil, as Figure 7.1 shows. The mother made no attempt to intervene, instead watching from the verandah with her four remaining kittens. Whether JB's actions reflected a need to maintain hygiene in the mother's bed or whether he grasped death's finality seemed irrelevant that day. With the mother's consent, JB had performed a ritual that seemed to satisfy both cat and dog.

In 1991, philosopher Michael Leahy argued that only animals with language can have an understanding and awareness of death, and only humans possess language (Humphreys, 2008). Based on this premise, he contended that animals on the way to be slaughtered do not suffer, nor do they suffer when others of their cohort are slaughtered in their view because they have no reflective understanding of death. The flaw in this argument is that there is a vast difference between practical awareness and reflective understanding, with evidence that animals can know something is about to happen without reflecting on it. My dog knows when we turn in the direction of the veterinary hospital and he does not like it. He does not need to reflect on why he does not like it, nor why he is going to the vet or what is likely to happen. He has an unpleasant feeling and so he resists it.

Other writers joined the debate on the ethics of killing animals based on whether the fact that animals cannot understand death should affect the extent to which death harms them. The argument that if animals have no concept of death, a painless death would do no harm was countered with the argument that death can still be harmful in the absence of concept of death. Philosopher Susana Monsó (2022) contended that death is ubiquitous among wild animals, making it tangible and present, as opposed to

Figure 7.1 Johnny Belinda remains close to the kitten he has just buried. Photograph by the author.

many humans who are shielded from the death of both humans and animals. Animals must recognise and respond to death within their social setting for survival reasons, including the presence of a predator and the need to replace by reproducing again, suggesting that they must have developed some understanding of the concept of death. It is essential to distinguish between sleeping offspring and dead offspring, or sleeping predator and dead predator, so as to inform behavioural choices related to care or avoidance (Brooks Pribac, 2021). JB chose the dead kitten, leaving behind the four living kittens with their mother. This discrimination suggests that animals can understand death as permanent and therefore requiring behavioural choices, in some cases prompted by the psycho-neurobiological components of grief identified in humans and some animals. When JB removed the dead kitten from his mother's bed, it could be inferred he had some understanding of the meaning of death and its irreversibility. Needing to touch and bury the kitten, a similar response to that shown by many social animals who sniff, touch, nudge and sit with corpses, suggested he was responding at an experiential level rather than a cognitive appraisal of the circumstances and outcomes.

For social animals, death does not belong to the individual alone as one death can have far-reaching implications for other members of the community. Death of an elephant matriarch, for example, results in loss of memories and is potentially detrimental to the whole group. Testing for understanding of death in a laboratory setting raises ethical concerns, while information from observational studies such as those cited earlier lack the scientific rigour deemed necessary for credibility. It is perplexing as to why some people have such difficulty in accepting that animals could have a concept of death, until one considers the ethical and moral implications inherent in such an admission when implicitly condoning animal mass death practices. The belief that death cannot harm an individual if that individual is not aware of their own mortality is premised on the anthropocentric assumption that the only way to conceptualise death is the human way and the only way to emotionally react to death is the human way. Seeking human-like reactions to death is doomed to fail, especially when humans themselves have different responses and understandings of death. Many human children do not cognitively understand the implications of death and its permanency until around five to seven years of age (Child Bereavement UK, 2022), but that does not mean that they lack emotional understanding of its existence several years earlier. This adds support to death awareness being on a continuum rather than binary, and therefore feasibly within the grasp of some animals. The following story of Irma, one of three bonded chickens, suggests that death awareness and the resultant grief is a very real possibility when chickens form strong emotional bonds with each other.

CASE STUDY–THE STORY OF THREE BONDED SISTERS

In September 2022, three chickens had the good fortune to arrive at Lefty's Place Farm Sanctuary in the Macedon Ranges of Victoria, Australia. Life prior to this turn of events had not been so fortunate for the three bonded sisters. Old, riddled with reproductive tumours and no longer laying eggs, the chickens were dumped in the bush to fend for themselves. Rescue and rehoming at Lefty's Place gave them the opportunity to learn that they were important, unique and loved, not just commodities in the service of humans.

Each animal who enters Lefty's Place Farm Sanctuary becomes an individual with a name, a personality and a story. That story often starts in the hidden suffering of factory farms and slaughterhouses, but changes dramatically with rescue, recovery and a new life at the sanctuary among friends. While the three hens were living on borrowed time due to the reproductive tumours so common with relentless egg laying, that time would be comfortable and safe. Named Marion, Miranda and Irma, the three sisters' story as they transitioned into unique little personalities was a favourite on social media, until a heart-breaking Facebook post appeared on 15 January 2023.

Marion, Miranda and Irma

"Marion left the world on Thursday after quite a few weeks on intense pain relief. She didn't give up or tell me she wanted to go, she didn't want to leave her two sisters, they were so incredibly bonded… So bonded that her sister Irma started vocalising in a way I'd never heard her vocalise before when she saw Marion's body".

Before long, Miranda also succumbed to the unrelenting tumours. Miranda was the leader and protector of her two sisters, always making sure Marion and Irma were safe. Once again, Irma was confronted with death, as Miranda's preparation to leave was described on social media:

"Miranda was brought in to a comfortable bed last night, given multiple pain relief, fluids and delicious food, but I knew she was on her way. This morning she was still hanging on. She was comfortable and not distressed but she didn't want to leave. I decided to bring her sister Irma in to spend her last hours with her. Irma stood in front of her and spoke to her. Within minutes of Miranda seeing and hearing Irma, she quietly passed away. She was waiting to say goodbye to her sister. I am so incredibly sure of it. That beautiful hen always wanted to make sure her sisters knew they were looked after and were going to be ok. Obviously she wanted to make sure Irma knew that she was leaving and she'd see her soon".

Irma watched as Miranda was buried with Marion, circling the grave and vocalising the whole time. A week later she could be found in places where the three sisters foraged together, looking around and calling to nobody.

Reproduced with permission from Lefty's Place Farm Sanctuary.

Humans control the life and death of so many animals, including whether it is a good death or a bad death. In their previous life, the death of three non-productive egg-layers would have gone unnoticed in the bushland to which they had been banished. The chicken who would later become Irma may not have had the luxury of saying goodbye to her dead sisters as she struggled to survive. In the safety of the sanctuary, Irma not only showed an awareness of death but also a response that in humans would be labelled as grief and bereavement. To acknowledge that a chicken has the emotional capacity to recognise and feel death is to invite cognitive dissonance at the treatment of countless anonymous chickens not so fortunate as the late Marion and Miranda and their surviving sister, Irma.

The lack of concrete evidence for an animal's awareness and understanding of death does not mean it does not exist, as evidenced by anecdotal stories of a cat who could predict death.

OSCAR THE CAT

On 26 July 2007, geriatrician Dr David Dosa's report about Oscar the cat appeared in *The New England Journal of Medicine*. Born in 2005 and adopted by staff members on the third floor of Steere House Nursing and Rehabilitation Center in Providence, Rhode Island, the article claimed that Oscar had predicted and presided over the deaths of more than 25 residents. The ability of humans to predict end-of-life in medicine comes with considerable perplexity and debate. Incorrect predictions can result in unwarranted or inadequate interventions, as well as alarmist or delayed notifications to family and friends. The possibility that a cat could get it right every time was the antithesis of modern healthcare setting protocols, and yet staff and families alike came to rely on the white and tabby cat's indefatigable sense of death.

According to Dr Dosa, Oscar's mere presence on a resident's bed prompted physicians and nursing home staff to notify family members of their loved one's impending death. Where no support was available, Oscar provided companionship to ensure no one died alone, an invaluable gift during the COVID-19 lockdowns of 2020. Oscar was subsequently immortalised in the *New York Times'* best seller book *Making Rounds with Oscar: The Extraordinary Gift of an Ordinary Cat* (2009). Acknowledging a range of physiological reasons for Oscar's behaviour (for example, Oscar's potential ability to scent biochemicals released by disintegrating ketone cells in the terminally ill) in no way lessened the compassionate comfort perceived by family members and staff from this gentle cat who, in his cat-like way, seemed to recognise death. Once the patient died, Oscar would leave the room, showing no human-like grief symptoms. When thinking back to the ants who recognise their dead conspecifics by the smell of triglycerides, it could be easy to dismiss these actions as an instinctive response to wanting to remove death from the home—except Oscar's behaviours also included perceived comfort and affection while the dying person was still alive.

On 22 February 2022, the Steere House Nursing and Rehabilitation Centre posted a brief message: "Oscar has passed away quietly."

CONTINUUM OF DEATH

Death represents humanity's greatest control over animals and often follows a series of atrocities inflicted on the animal for human benefit. Humans have the power to bring animals into life and manipulate their existence, with the sole intention of ultimately killing them. This should result in cognitive dissonance at an unbearable level, and yet billions of animal deaths remain categorised as necessary, humane or simply not to be thought about. Whether animals have an awareness of their own death must therefore remain a fuzzy grey area if people are to continue the mass destruction of animals with an easy conscience.

Ironically, an animal's ability to recognise and respond to death is implicitly acknowledged in some animal research ethical standards, such as the *Australian Code for the Care and Use of Animals for Scientific Purposes* (the Code; National Health and

Medical Research Council, 2013). Item 3.3.45 specifies that methods and procedures for killing an animal must be humane, and "ensure that animals are killed in a quiet, clean environment away from other animals". These standards specify training for personnel tasked with killing the animals and methods to ensure an ethical death (such as CO_2 with specified percentages, or cervical dislocation among small rodents). The location of death for mammals such as rodents must be in an area separate to other animals so as to prevent stressful auditory, visual and olfactory stimulation of the remaining cohort. While this can be interpreted as an acknowledgement that these mammals have a rudimentary death awareness, it may also be to avoid the release of stress hormones prior to death that could potentially skew the study results.

Standards also govern the use of reptiles and fish in research but a separate location of death is not always specified, perhaps reflecting human perceptions of reduced awareness and emotional capacity in non-mammals. Legal protection for insects in research is lacking in many countries, despite increased evidence that insects possess faculties conducive to emotive states and self-awareness. In some cases, definitions of what constitutes an animal in research stop at the higher order invertebrates such as cephalopods, leaving all taxonomically lower species as vulnerable as the cats that my father rescued in the 1960s. These different approaches to death for mammals, reptiles, fish and invertebrates in bioethics and science policy may be partly driven by stereotypes about the rigid instinctual behaviours of some groups to the exclusion of any potential for emotion and death awareness.

Moving further along the continuum of death are the farmed animals (including edible insects farming) perceived as emotionless commodities and destined for a largely un-mourned mass death. Works such as Kathryn Gillespie's (2018) *The Cow with Ear Tag #1389* give a much-needed voice to these sentient creatures in the face of statistics that millions of cows are slaughtered annually for meat. When asked if cattle are aware that they will die, Dr Temple Grandin (2001), designer of livestock handling facilities and a Professor of Animal Science, commented that she believed cattle did not understand they would be slaughtered. This was based on no discernable behavioural differences between cattle at slaughter plants and cattle in feedlot veterinary chutes, and the expectation that they would be more agitated at a slaughter yard if they understood death and knew they were about to die.

Temple Grandin's website offers a wealth of information on livestock behaviour, design of facilities and humane slaughter based on her decades of experience and research. However, the overall impression is slightly marred by the descriptor attached to the humane slaughter section and preceding recommended stunning practices. Death for animals destined for human consumption takes on an anthropocentric perspective in terms of the human palate, not just animal welfare with the advice that:

"Stunning an animal correctly will provide better meat quality. Improper electric stunning will cause bloodspots in the meat and bone fractures. Good stunning practices are also required so that a plant will be in compliance with the *Humane Slaughter Act* and for animal welfare. When stunning is done correctly, the animal feels no pain and it becomes instantly unconscious".

Consumption of meat has grown exponentially over the last century, making modern farming methods essential to keep up with demand. Initiatives to ensure animals do not suffer, or suffer minimally, at point of death may allay cognitive dissonance, but fail to address the events leading up to their death and the unavoidable fact that they are dying for human needs.

HIERARCHY OF DEATH

Over a decade ago, I learned of the death of a puppy in Darwin. The naïve six-months old had gently mouthed a cane toad in an invitation to play and was dead within hours. Cane toads were introduced to Australia in 1935 in a desperate attempt to eradicate the native cane beetles that were destroying sugarcane crops in Queensland (West, 2018). Unfortunately, the 100 toads released started devouring anything they could fit in their wide mouths, including rodents, birds and household rubbish, while reproducing rapidly. Highly toxic and with no natural predators, by 2021 there were an estimated 200 million cane toads in the northern and eastern parts of Australia forcing them to compete for resources against each other. Australian cane toad tadpoles solved this problem through cannibalisation of hatchlings (Kozlov, 2021).

Over one hundred cane toads died to avenge the puppy's death when the grief-stricken human family retaliated with spades and golf clubs. Cane toads are the awkward creatures with looks and actions so alien that they have forfeited all rights to a humane death. Positioning the cane toad as an alien invader in Australia legitimises slow and painful methods of killing—such as bludgeoning, suffocating in plastic bags or pouring chemicals on them—by adults and children alike. These methods would not be tolerated in research facility amphibians or among other amphibians perceived as valuable to the environment, less poisonous and less ugly. The cane toad is undoubtedly wreaking havoc on the natural environment and is proving a formidable adversary, but their manner of death has become a social construct determined by the animal's positioning as an ugly, deadly, cannibalistic foe. Devoid of any perceived emotional capacity other than greed and evilness, the animal's capacity to feel pain and suffer prior to their human-inflicted death is forfeited in the battle to reclaim the land for native wildlife.

Applying the Zoological Emotional Scale to cane toads allows a visual representation of this amphibian's position in the environment to which it was intentionally introduced by humans. This provides a tool to understand and interrogate the numerous perspectives on cane toads and their capacity to have feelings beyond being a voracious puppy killer. There is no denying that cane toads have negatively impacted their non-natural environment but encouraging children to kill these creatures serves to solidify human domination over the life and death of animals and obscures consideration of other means of controlling a human-made problem. When applying this framework, each person starts from their own point of familiarity with the animal, ranging from first-hand experience of a puppy killer to learned information about the resilient amphibian's relentless spread throughout Australia.

Figure 7.2 represents the outcomes of some of the possible interactions between cane toads and humans in Australia and the subsequent positioning in each domain of the Zoological Emotional Scale. Cane toads have a function in their natural environments of South and Central America, thus demonstrating that treatment of an animal can be dependent on geographic location.

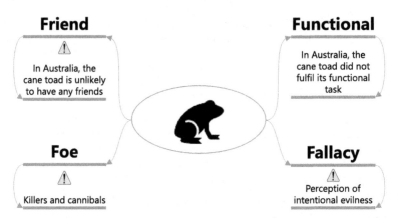

Friend

⚠

In Australia, the cane toad is unlikely to have any friends

Functional

In Australia, the cane toad did not fulfil its functional task

Foe

⚠

Killers and cannibals

Fallacy

⚠

Perception of intentional evilness

Figure 7.2 The Zoological Emotional Scale applied to the cane toad indicates it is unlikely that any emotional capacity is attributed to this amphibian.

IMPLICATIONS

Humans have such a complex relationship with death that it seems inevitable this attitude should also taint beliefs around animal awareness of death. Pervasive to most scenarios of animal death is the power of humans to control if, when and how an animal should die. This is accompanied by the need in some cases to allay cognitive dissonance through the provision of a perceived good death that legitimises the right to kill other sentient creatures in massive numbers. Fundamental to a good death is belief in an animal's inability to cognitively appraise the meaning of death, a belief that is increasingly being challenged through observation and anecdote.

However, many urban dwellers have very little first-hand experience of animal death on a commercial scale. The neatly packaged outcomes of animal mass deaths through factory farming are far removed from the bloody killing process, just as the vaccines and other medicines are disconnected from the research animals who died. Fundamental to the action phase of any change process is understanding why certain practices related to animal death are perpetuated without question. The Zoological Emotional Scale invites questioning around these practices, the implicit condoning of mass death and the mechanisms that keep the discomfort of cognitive dissonance at bay.

Experiences of death for many urban dwellers can be limited to companion animals as they age or sicken and it is here that sharing or relinquishing ultimate control over individual animal deaths can provide a catalyst to change. Much of the literature around euthanasia of companion animals focuses on the human carer, their guilt and their grief, as well as the emotional toll this takes on the veterinary professionals as the providers of death. While these are important and valuable areas of consideration, the animal is also an integral stake-holder in the decision-making process. Human control in this process is driven by the implicit assumption that animals cannot speak for themselves or engage in their own care, despite ever-evolving literature on the volitional capacities and agency of animals. Jessica Pierce (2019) draws attention to the need to recognise and find ways in end-of-life care to respect consent, assent and dissent of the animal themselves rather than adopting a paternalistic approach that fails to accommodate the rights and abilities of the animal.

Changing this perspective to incorporate the animal's autonomy as an emotional being provides another means by which to acknowledge and respect an animal's innate awareness of death. When Bella, the matriarch of my cats experienced Feline Aortic Thromboembolism, euthanasia seemed a logical outcome. Placing Bella on the floor, the vet and I stood back and simply watched as she glared at us and struggled to rise. Despite the paralysis, her behaviour, her emotional response and her defiant attitude suggested she was not ready to die and so we complied with her decision.

REFERENCES

Atwal, S. (2022, January 12). *190-year-old Jonathan becomes world's oldest tortoise ever.* Guinness World Records. https://www.guinnessworldrecords.com/news/2022/1/190-year-old-jonathan-becomes-worlds-oldest-tortoise-ever-688683

Bekoff, M. (2007). *The emotional lives of animals: A leading scientist explores animal joy, sorrow, and empathy--and why they matter.* New World Library.

Brooks Pribac, T. (2021). *Enter the animal: Cross-species perspectives on grief and spirituality.* Sydney University Press.

Child Bereavement UK (2022). *Children's understanding of death at different ages.* https://www.child-bereavementuk.org/information-childrens-understanding-of-death

Cowgill, V. M. (1972). Death in Perodicticus. *Primates, 13,* 251–256.

Dosa, D. (2007, July 26). A day in the life of Oscar the cat. *New England Journal of Medicine, 357*(4), 328–329. https;//doi.org/10.1056/NEJMp078108

Gillespie, K. (2018). *The cow with ear tag #1389.* University of Chicago Press

Grandin, T. (2001). Welfare of cattle during slaughter and the prevention of nonambulatory (downer) cattle. *Journal of the American Veterinary Medical Association, 219,* 1377–1382.

Greenberg, J. (2012). Terror management theory: From genesis to revelations. In P. R. Shaver & M. Mikulincer (Eds.), *Meaning, mortality, and choice: The social psychology of existential concerns* (pp. 17–35). American Psychological Association. https://doi.org/10.1037/13748-001

Griffin, D. (1981). *The question of animal awareness: Evolutionary continuity of mental experience* (Revised and enlarged edition). William Kaufmann, Inc.

Humphreys, R. (2008). Animal thoughts on factory farms: Michael Leahy, language and awareness of death. *Between the Species.* Issue VIII. www.cla.calpoly.edu/bts/

Kozlov, M. (2021, August 25). Australia's cane toads evolved to be cannibals at frightening speed. *Nature, 597,* 19–20. https://www.nature.com/articles/d41586-021-02317-9

Langer, S. K. (1962). *Philosophical sketches.* Johns Hopkins Press.

Langs, R. (2004). Death anxiety and the emotion-processing mind. *Psychoanalytic Psychology, 21*(1), 21–53. https://doi.org/10.1037/0736-9735.21.1.31

Miller, G. A., Galanter, E., & Pribram, K. H. (1960). *Plans and the structure of behavior.* Holt, Rinehart & Winston.

Monsó, S. (2022). How to tell if animals can understand death. *Erkenntnis, 87,* 117–136. https://doi.org/10.1007/s10670-019-00187-2

Monsó, S. & Osuna-Mascaró, A.J. (2021). Death is common, so is understanding it: the concept of death in other species. *Synthese, 199,* 2251–2275. https://doi.org/10.1007/s11229-020-02882-y

National Health and Medical Research Council (2013). *Australian code for the care and use of animals for scientific purposes,* 8th edition. National Health and Medical Research Council.

Pierce, J. (2019). The animal as patient. Ethology and end-of life care. *The Veterinary Clinics of North America. Small Animal Practice, 49,* 417–429. https://doi.org/10.1016/j.cvsm.2019.01.009

Safina, C. (2020). *Becoming wild. How animals learn to be animals.* Oneworld Publications.

Temple Grandin's Website (n.d.). Livestock Behaviour, Design of Facilities and Humane Slaughter. Retrieved from https://www.grandin.com/index.html

Uy, C.A. & Johnson, D.W. (2022). Effects of microplastics on the feeding rates of larvae of a coastal fish: Direct consumption, trophic transfer, and effects on growth and survival. *Marine Biology, 169*(27). https://doi.org/10.1007/s00227-021-04010-x

West, P. (2018). *Guide to introduced pest animals in Australia.* CSIRO Publishing.

Yalom, I.D. (1980). *Existential psychotherapy.* Basic Books.

Yan Wang, Z. & Ragsdale, C.W. (2018). Multiple optic gland signaling pathways implicated in octopus maternal behaviors and death. *Journal of Experimental Biology, 221*(Pt 19). https://doi.org/10.1242/jeb.185751

SPIRITUALITY AND EXPERIENTIAL CONSCIOUSNESS

8

On warm evenings over the summer months, my two elderly dogs can be found quietly sitting together on the back verandah facing due west as the sun drops behind the massive Jacaranda tree. One completely blind and the other slowed with arthritis, they maintain this routine that started years earlier with their deceased predecessor. Mindful of the temptation to anthropomorphise a scenario that may in fact be two dogs digesting their dinner quietly, their behaviour also suggests experiential enjoyment of a peaceful, almost sacred time of contemplation for these lifelong companions.

This practice began over 15 years ago with two other dogs who would routinely head out to the back verandah after their meal as the sun disappeared behind a much smaller Jacaranda tree. The death of one dog left a gap that was eventually filled with an anxious little rescue dog who could not stop fidgeting as he sat slightly behind his elderly mentor in Figure 8.1. The night before the older dog succumbed to his multiple ailments, I can still see them sitting together as the sun went down one last time in the old dog's life. Eyes closed, while the younger one remained alert, the ritual appeared to provide comfort for both. Again, aware of the risk of anthropomorphising this tranquil scenario, it nevertheless caused me to stop contemplating the imminent death of my best friend and instead embrace the experiential moment provided by watching these two dogs.

Watching the sunset in the same spot remained a habit with the fidgety rescue dog and the youngster who wandered into our lives several months later. There is something about that precise spot on the back verandah at the precise moment the sun disappears that triggers a sense of calmness and peace in the now quite elderly dogs. Whether this is awe and spiritual engagement remains irrelevant as I watch my two old friends and quietly remember the two who are missing. It does not need complex thought, nor does it need language, to perceive the positive affective state that my dogs and I derive from this ritual.

This chapter introduces the concept of spirituality (not religiosity) as part of an animal's experiential consciousness. Describing animals as spiritual beings can be controversial as it encroaches on a domain jealously guarded by some humans for centuries. The historical origins of spirituality as "life in spirit" remained in constant use in Western nations until around the twelfth century when there was a slow move towards the concept of spirituality as a way of distinguishing intelligent humanity from any non-rational creations, such as animals (Sheldrake, 2014). With roots in the

DOI: 10.1201/9781003298489-9

Figure 8.1 Two dogs lie together in their evening ritual, one with closed eyes and the smaller dog alert but unsure where to look. Photograph by the author.

Latin words *spiritus* (meaning breath, courage, vigour or soul) and *spirare* (meaning to breathe), spirituality has been linked to five characteristics: meaning, value, transcendence, connecting (with oneself, others, God or supreme power and the environment) and becoming (growth in life) (Ghaderi et al., 2018). Therefore, at a fundamental level, there is nothing in this definition that specifically excludes nonhuman animals from a practice that creates a sense of peace, purpose and connection with others. However, in contemporary society where many animals are disconnected from members of their own species, other living creatures and their natural environment, human domination has diminished opportunities for realisation of an innate spirituality.

Among humans, three broad approaches to spirituality have been identified: classic religious spiritualities; esoteric spiritualities combining religious elements with philosophical or ethical ones; and increasingly, non-religious or secular understandings of spirituality. It is this latter approach that encompasses ecologically sensitive spiritualities that form the basis of this chapter. Much of the research about spirituality relates to humans and their quest to connect with themselves and a greater being (often nature). The Biophilia Hypothesis describes humanity's innate need to connect with nature, while one broadly accepted contemporary definition of spirituality as "[...] the aspect of humanity that refers to the way individuals seek and express meaning and purpose and the way they experience their connectedness to the moment, to self, to others, to nature, and to the significant or sacred" (Puchalski et al., 2014, p. 643) refers specifically to humanity but positions nature as essential to its fulfilment. In these definitions, animals are a conduit to human spirituality without being privy to the end experience. They facilitate an awareness of the interconnectedness of all life, making them an essential component of human spirituality and the need to live in harmony with and respect the natural environment. As such, they can allow humans to transcend the mundane and enter an expanded level of consciousness that in turn facilitates compassion, care and understanding for all species (Faver, 2009).

However, there remains a subtle demarcation that risks turning these definitions with claims of interconnection into anthropocentric affirmations of human superiority. Animals are positioned as commodities to allow humans to achieve a level of consciousness that the animals can never hope to achieve. Restricting animals and nature to the periphery of human spirituality reflects implicit prejudices, especially evident in Western psychology which largely omits any consideration of spiritual states of consciousness in animals (Cunningham, 2022). This prejudiced perception has also contributed to the silencing in modern industrialised societies of indigenous voices that for millennia have positioned animals as spiritual creatures.

While an awareness of the interdependence of all life and the need to protect the earth may be embedded within human spirituality, this view is not always adhered to in parts of the world where urbanisation, industrialisation, abuse and exploitation of animals for human needs dominate. Spirituality must be denied animals as to do otherwise risks cognitive dissonance over the suffering inflicted on these sentient beings. Taking a more eco-centric perspective such as that encompassed within eco-spirituality would require acceptance not only of the need for spiritual connection between humans and the environment, but demonstrated actions to recognise and nurture animals as equivalent partners in this connection.

Literature related to childhood spirituality comes closer to providing the foundation for an animal-inclusive approach. Childhood spirituality has been linked to the strong presence of creative thought, play, wonder and sheer joy (Sheldrake, 2014). An important key to a child's spirituality lies in the priority of emotional sensitivity over intellectual reasoning, which facilitates the development of a perception of the sacred in the layered meanings of their story worlds. The butterfly in Figure 8.2 provides an

Figure 8.2 A butterfly with open wings rests on a leaf. Photograph supplied by the author.

example of the child's ability to gaze in awe at the beauty of a living creature without ascribing any other meaning than enjoyment in the moment. This assumes the child has seen a living butterfly, as urbanisation increasingly removes access to many forms of natural life. For adults, the relationship can be more complex, ranging from the butterfly as a potential pest whose caterpillar offspring munch through urban patches of greenery to embracing one of the more attractive insects that do not prompt revulsion when providing a conduit to nature. Neither perception respects the butterfly as a living creature in its own right and a significant contributing member of the natural environment.

When exploring spirituality in animals, Marc Bekoff (2001) proposed animal play as both Sacred (trust in mutual agreement) and Soulful (being at the moment from deep in their hearts) and therefore key components of animal spirituality. Being at the moment is reminiscent of the concept of mindfulness, a meditative process that focuses a person's cognitive, attitudinal and affective attention and awareness on acceptance of the present moment. Animal-based mindfulness training has attracted some research attention, again based on the positioning of animals as conduits to mindful awareness of the supportive contemplative effects of the natural world. The persistent emphasis on animals as conduits but not spiritual equals demonstrates human perceptions of exceptionalism that fails to acknowledge a shared journey by spiritual equals. The *Cambridge Declaration on Consciousness* confirmed that many animals do have the neural capacity to focus their affective attention, thus shifting them from adjuncts of the human experience to independent players capable of sitting quietly with close others as they watch a sunset or some other awe-inspiring landscape.

CASE STUDIES – THE WONDER OF NATURE

Jane Goodall created controversy when pondering the "waterfall dances" of chimpanzees as possibly a joyous response to being alive, or perhaps an expression of awe at the wonders of nature (Goodall & Berman, 1999). She questioned:

> For ten minutes or more they may perform this magnificent "dance". Why? Is it not possible that the chimpanzees are responding to some feeling like awe? A feeling generated by the mystery of the water; water that seems alive, always rushing past yet never going, always the same yet ever different. Was it perhaps similar feelings of awe that gave rise to the first animistic religions, the worship of elements and the mysteries of nature over which there was no control?
>
> (p. 188)

The interpretation of spirituality among these chimpanzees has been disputed, with Christopher Fisher (2005) commenting that this event cannot be symbolic in the way humans understand something to be symbolic, as there is no meaning without language. He refuted that animals had the capacity for interpreted experience of meaningful spirituality of equivalence to humans, instead equating the emotive descriptions to anthropomorphism. This sentiment is echoed in the teachings of philosophers such as Hegel, Kant and Plato who proposed that accomplishments in moral reasoning and linguistic capacities set humans above other animals and removed them from any semblance of shared life at a spiritual level (Willett, 2014).

Building on the waterfall experience and decades of chimpanzee research, Jane Goodall provided an intriguing entry in the *Encyclopedia of Religion and Nature* entitled "Primate Spirituality" (Taylor, 2005). After describing the chimpanzees' foot thumping, rock throwing, jumping and hooting followed by quiet sitting and staring behaviours, Goodall suggested that these behaviours and emotions may demonstrate the origins of spirituality in human primates and "precursors of religious ritual" (p. 1304). Within an evolutionary perspective, it made sense that the qualities of consciousness, self-awareness, emotion, moral sense and imagination that contribute to spirituality in humans did not just appear without some form of prior evolutionary refinement. Charles Darwin himself emphasised the continuity of phylogenetic, affective, behavioural and psychological processes among species, adding further weight to the existence of animal spirituality.

Jane Goodall's waterfall example is not an isolated incident. Anthropologist Barbara Smuts described coming on a troupe of baboons in Tanzania, sitting paused and motionless by a stream's still pools (Willett, 2014). Alone or in small groups, even the noisy juveniles sat in silent contemplation until some unseen signal saw them resume their journey. Smuts surmised that the still waters seemed to unlock a sacred experience of oneness with nature, very much like meditation, thus hinting at animal capacities for peacefully transposing ordinary life into an altered state of spirituality. When describing Smuts' experience, philosopher Cynthia Willett (2014) asked:

> Wouldn't it be funny if the ability to immerse oneself in the flow of life, that is to live for a moment fully in the waves of the present—a state thought to determine animal modes of awareness as inferior to human consciousness—is in fact not the lowest stage of consciousness but the highest?
>
> (p. 130)

Ethologist Marc Bekoff (2001) commented that while science can, and has offered many answers over the centuries, it does not hold a monopoly on truth. A lack of knowledge or evidence does not mean something cannot or does not exist, a timely reminder when thinking of the history and controversy surrounding the possibility of emotional capacity, and even the ability to feel pain, among animals. Bekoff reiterates: "Science also cannot prove that animals have deep emotional lives or souls, but it also cannot prove that they do not. Often science discounts possibilities on the absence of data" (p. 619). Given the lack of scientific consensus about spirituality in humans (who can communicate in a language understood by scientists), it seems inevitable that consensus about spirituality in animals remains divided. Nevertheless, there is increasing anecdotal evidence of some animals' capacity to live and enjoy the moment, as Marc Bekoff described with animals that sometimes "just go nuts". There was the young black rhinoceros in Kenya who ran all over the place with joyful abandon when there were no other playmates. At Edgar's Mission sanctuary for rescued farmed animals in Australia, Fifi the sheep indulges in solitary "sunset zoomies" in the paddock against a backdrop of the setting sun, while Gracie and Dottie, two rescued cows, approach sunset in a different mystical way.

Gracie and Dottie, as Told by Edgar's Mission

And so as we watch dear Gracie and Dottie, the warm sun's rays caressing us all on this glorious Autumnal day, a certain mysticism fills the air that one cannot escape. Inhaling deep, it floods one from head to toe, and we are reminded that it is indeed the same sun, air and water that fuels us and they. We have the same need for food, shelter and company, and we too cherish our loved ones.

We watch a little more and see them savour the fodder before them and realise that this too is fine fodder for us to do better, to think clearer and to live kinder for their kind. For when we do this good fortune shall shine on all…

Reproduced with permission from Edgar's Mission.

Jane Goodall's chimpanzees, Barbara Smuts' baboons, Marc Bekoff's black rhinoc-eros, Gracie and Dottie, and my own two elderly dogs fit neatly into professor of the-ology, culture and spirituality Lawrence Cunningham's (2022) working definition of animal spirituality as "any experience that increases the meaning, purpose, or quality of whatever life an organism feels to be at the centre of its being and promotes the organism's flourishing" (p. 193). Humans will never know what each of those animals was feeling as they gazed or frolicked, but something had drawn them to focus on an experience that appeared to have some meaning for them. This definition removes the need for language and interpretive cognitive functions, often cited to support human exceptionalism in this domain, and returns it to a non-reflective, emotional, relational experience between the animal and whatever has prompted these feelings. Each of these animals appeared to gain access to affective feelings in the here-and-now, a fac-tor that plays a central role in definitions of spirituality among humans. For example, the Puchalski et al. (2014) definition includes "[…] and the way they experience their connectedness to the moment, to self, to others, to nature, and to the significant and sacred". Because animals' cognitive capacities are perceived as not sophisticated enough for abstract thinking and the meaning-making associated with spirituality, its existence in animals has been contested or dismissed. This premise is fundamentally flawed, as activation of too much cognitive processing may, in fact, block spiritual experiences in the moment, suggesting that animals (and children) may be in a stronger position to achieve elevated feelings of wonder and awe than adult humans (Willett, 2014).

Once spirituality in animals is shifted away from the cognitive, interpretative domain more strongly linked to human religion and relocated in the feeling domain of animals (and humans are also animals), the possibility of animal spirituality becomes a reality. The animal kingdom is wide, and as with emotions, expressions of spirituality will differ between animal species and humans. Difference, however, does not preclude its exist-ence except where humans have suppressed this fragile quality within the animals they dominate. When Lawrence Cunningham defined animal spirituality, he emphasised experiences that increased the meaning, purpose or quality of whatever life resided at the centre of an animal's being and promoted flourishing. Gracie and Dottie, the cows at Edgar's Mission, flourished in feeling the Autumnal sun on their bodies. This allowed them to find their experiential selves away from the commercial farms that banish freedom and sunshine from the lives of so many cows. The privilege of communicating with nature and the environment, as described by Jane Goodall and Barbara Smuts, seemed to facilitate an experience akin to awe rather than driven by instinct.

While spiritual experiences, like emotional experiences, are more easily identified among the familiarity of mammals, Michael Tobias, environmentalist and filmmaker, described an intrinsic spirituality among the huge, endangered whale sharks with whom he swam: "The sharks exhibit bliss, the ultimate state of meditation and indwelling referred to by such diverse luminaries as Buddha and Thoreau" (Tobias, 2000, p. 173). Whale sharks do not eat humans, nor are they whales. Instead, they are gentle giants whose perceived playfulness and willingness to afford humans the opportunity to get up close to a wonder of nature may be their downfall. Despite having several hundred teeth and a mouth over 1.5 m wide, they are filter feeders who swim forward and swallow krill, crab, fish, larvae, small schooling fish and jellyfish. They are vital to maintaining a balanced ecosystem, but face dangers from propellor strikes and disrupted feeding when they become the focus of tourism and photo opportunities.

Figure 8.3 represents some of the perceptions humans hold around the emotional and spiritual capacity of the world's biggest fish. Positioned as a friendly gentle giant, reinforced by the belief that these sharks enjoy being a tourist attraction, visitors are tempted into the water by descriptors of the experience as awe-inspiring, exhilarating and life-changing. This raises the question of how many people perceive these animals as the spiritual beings described by Michael Tobias, or a conduit to a wildlife encounter as described by the tour operators.

Whether invertebrates, especially insects, experience anything akin to increased meaning, purpose and connection remains unfathomable. It may be beyond human comprehension to attribute anything other than instinct to these creatures, despite the research noting self-awareness in bees and self-sacrificing behaviours in eusocial insects such as ants and bees.

In human literature, spiritual awareness has been demonstrated through the expression of positive traits such as compassion, love, altruism and empathy (Huber & MacDonald, 2012). In fact, non-religious spiritual cognitions and experiences can be significant predictors of both empathy and altruism, two qualities that have been identified among many animals, including some invertebrates. Overcoming the tendency to perceive animals and nature as simply conduits to human spirituality may challenge some fundamental human principles of exceptionalism, succinctly summed up by Teya

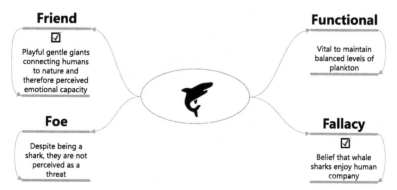

Figure 8.3 The Zoological Emotional Scale is applied to the whale shark to show where emotional capacity and spirituality are attributed to this large fish.

Brooks Pribac (2017) as "Spirituality is one such attribute, jealously guarded by those whose humanity may feel under threat if it turns out that the sacrificial lamb has equal, or greater, spiritual depth compared to her killer" (p. 356).

Blocking animal spirituality becomes a necessity if the sacrificial lamb is to be forfeited without guilt—or if the animal slaughterhouse worker and researcher are to continue dealing in animal death on a daily basis. Pondering for a moment the spiritual significance of the lives of millions of farmed animals and rodents bred to die would make the guilt and conflict of cognitive dissonance unsustainable.

CASE STUDY – THE SACRIFICIAL GOAT

Goats have a long history of sacrifice as part of human religious or cultural traditions. Although illegal in many countries, anecdotal reports suggest slitting the throats of terrified goats may still occur in some private areas. Sociologists Henri Hubert and Marcel Mauss (1981) in their 1898 seminal book *Sacrifice: Its Nature and Functions* stressed that the sacrificial victim did not come with a sacred or spiritual nature. Instead, the sacrificial ritual simultaneously conferred fleeting sacredness on the victim and resulted in its destruction. The ceremony, no matter how painful and distressing for the animal, was crucial in establishing a means of communication between the sacred and the profane human world, thus positioning the sacrificial goat's death as nothing more than a conduit for human transcendence.

The goat's ability to graze and browse on poor-quality forage and water-scarce land makes them popular with smallholder farmers in some developing nations. Providing meat, milk, manure, skin and hides, as well as playing a role in religious and cultural ceremonies, goats are traded in informal markets and slaughtered using unregulated traditional methods. As the commercial value of goats increased in developed nations, their welfare at slaughter came under the spotlight of the European Commission's Food Safety Authority (2021). The resulting 2021 report recommended that goats should not be kept in lairage (holding pens) for a period prior to slaughter and stipulated basic welfare considerations such as access to water if this was unavoidable. The report also identified cues to help determine if any goats were experiencing pain and fear. These cues included slipping, falling, escape attempts, vocalisations, injuries, reluctance to move and turning back. It was reiterated that painful handling such as lifting, dragging by the horns or one leg, hitting with a stick and the use of dogs should be avoided.

The need to provide a reminder of these common-sense welfare instructions and universally recognisable signs of pain and fear in a twenty-first century welfare report is testament to the ongoing perception of farmed animals as unfeeling commodities for human consumption. One purpose of lairage is to allow abattoirs to operate at a consistent speed by always having a supply of animals on hand, an implicit reminder that irrespective of any basic welfare considerations, the goats faced death once they arrived at the slaughterhouse.

The versatility and resilience of goats over the centuries has embedded them within an anthropocentric perspective of sacrifice for human spiritual transcendence or slaughter to meet a global demand for protein. Both viewpoints ignore the charismatic personalities of these animals and fail to respect the goat's right to life outside of human needs. To understand the goat as a unique individual, it is necessary to turn to the farmed animal rescue sanctuaries for a less biased perspective.

BUTTERCUPS SANCTUARY FOR GOATS

As goats increasingly join the ranks of mass-produced farmed animals, a small sanctuary in Kent, United Kingdom has been rescuing mistreated, abused or abandoned goats since 1989. Buttercups Sanctuary for Goats provides an environment of love and kindness where goats are recognised as individuals with names, unique personalities and amazing life stories.

Away from slaughter or sacrifice, goats have a zest for life, which can emerge in unpredictable and chaotic ways. The sanctuary's 2022 Highlights newsletter succinctly sums up the joy—and risks—of living with these charismatic and unique animals:

> Anybody who has spent any time around goats, will know that their natural element, is unbridled chaos. They break things and they steal things, they go under things and they go over things [...] You can be certain that not a day will pass, without some unforeseen goat-related mayhem cropping up.

Reproduced with permission from Buttercups Sanctuary for Goats.

For all their idiosyncrasies and mayhem, there is no doubt that these goats, many of whom started life in trauma, are respected as individuals and afforded the opportunity to enjoy the highs and lows of a full emotional life at Buttercups. One such character was Lucky, described as a trusting, affectionate, free spirit who was not averse to checking out an unsuspecting person's pocket in search of sweets. Lucky's story represents the worst in human cruelty. He was not a sacrifice nor destined for human consumption. He was simply abused—his tail cut off, his throat slit and thrown into a pond to drown—all by the age of one month.

Lucky joined the Buttercups Bachelor Boy herd in 2012, where he was free to create mayhem or simply chill out surrounded by his goat friends in a loving, caring and safe environment until his death in 2020. Drawing on Lawrence Cunningham's (2022) definition of animal spirituality as those experiences that promote flourishing and increase the meaning and purpose of what an animal feels to be at the centre of their being, for eight years Lucky arguably achieved greater spiritual depth than the humans who had abused him. No longer a disposable commodity, he was a unique creature whose life and death had held meaning after arriving at Buttercups Sanctuary.

Millions of animals globally are deprived of their natural environment and behaviours as they become commodities whose life purpose is to provide food, products or entertainment for humans. Any semblance of a spiritual and meaningful existence is extinguished for those animals raised in restricted spaces, never seeing sunlight, forced to behave in ways contrary to their nature and deprived of any experiences with meaning, purpose or quality that could promote flourishing. Melanie Challenger (2021) commented that any spiritual connection between humans and animals cannot be mutual as long as the world remains dominated by human animals who do not think they are animals. Released from this domination and provided the opportunity and freedom to be themselves, many animals like Lucky demonstrate joy at being alive in the moment, a qualitatively different experience to simply surviving. Human language may be absent, but the fundamentals of a spiritual presence, including self-awareness, meaning-making and the ability to live in the moment, are present. Depriving animals of access to their natural world and control over their lives can quickly destroy the spiritual light in their eyes, as the history of dancing sloth bears poignantly demonstrates.

WHEN THE SPIRITUAL LIGHT GOES OUT

Imagine seeing your mother shot as you are brutally snatched, sold on the black market and forced to dance in a parody of light-hearted *joie de vie*. Before hitting the roads bedecked in ribbons and bells, you must learn to dance on command. First, your teeth must be filed down or knocked out for the safety of those watching you dance. Your claws must be extracted, and a metal ring or a thick, coarse rope threaded through your sensitive muzzle. As if that is not enough, you must now learn to dance by being forced to stand on hot metal sheets, hopping from foot to foot to relieve the burning pain. As you jerk around in a sick parody of dancing, music plays and before long this behavioural conditioning sees you performing every time you hear the music and your nose rope tugs. There was no physical, emotional or spiritual joy in being alive for the "happy" dancing bears of Asia.

Dancing bears embodied a travesty of animal emotion and spirituality. For centuries, the sloth bear provided cheap roadside entertainment in parts of the Indian subcontinent before being banned in India in the 1970s. Some bears continued to be used illegally as wildlife agencies worked tirelessly to rescue and rehabilitate them and provide alternative sources of income for the humans. Ranglia, one of the last dancing bears to be rescued in 2017, was eventually relocated to Wildlife SOS Agra Bear Rescue Facility in India, where he continued to exhibit signs of psychological stress and trauma (Sen Nag, 2020). As Ranglia's joy in life returned, he slowly learned to explore his environment without being tied to a rope, play in a meaningful sloth-like manner and take naps in mud pits he could dig for himself. Previous displays of human-centric false happiness gave way to sloth bear-centric true happiness with freedom and re-connection to a natural environment. Lying in a mud pit of his own creation, fully aware of the feel of dirt and sunshine at a fundamental, experiential level, is no different to Gracie and Dottie's enjoyment of the Autumnal sun and very much what humans might call a spiritual experience in sloth bear style.

The spiritual light continues to be extinguished in the many wild animals used and abused for human entertainment. From tigers forced to run, jump, swim and beg on command to koala bears who must endure close cuddling and photo shoots with humans, the burden imposed on animals to keep humans happy is immense. Biophilia Hypothesis describes human's innate need to connect with nature, while eco-spirituality describes the manifestation of the spiritual connection between human beings and the environment. Where the animal's spiritual light has been extinguished, their capacity to provide humans with a conduit to spirituality becomes a charade at the expense of their wellbeing.

IMPLICATIONS

Respecting animals as spiritual beings in their own right can encroach on areas deemed exclusively human in some cultures. Co-existing with animals, rather than having dominion over them, can be hindered by a reluctance to accept and respect animals' capacity to feel, an important component in their spiritual existence. However, cognitive dissonance remains a risk when acknowledging some animals' rights to a spiritual existence, while depriving those whose bodies or behaviours serve humanity in alternate ways. Visit an animal research laboratory and you will find mice and rats stacked in boxes on shelves where the inhabitants will never feel sunshine, fresh air or any semblance of a natural environment in their short lives. Observe factory-farmed chickens confined to cages and never able to forage or even touch solid ground and it becomes evident that humans can withhold the basic ingredient of nature that is fundamental to spirituality.

Similarly, positioning animals as simply a conduit to human spirituality is another use of animals with no recognition of their own right to this concept. There is an implicit hierarchy as to which animals are conduits, with a butterfly offering greater scope than a cockroach. A young child is taught to marvel at the fragile beauty of a butterfly but through social learning will recoil in horror at the resilient speedy cockroach. Both play a role in ecosystems but the cockroach's greater adaptation to human life has seen them vilified as harbingers of disease rather than a conduit to the natural world. Ironically, the potential loss of biodiversity and extinction of many insects will see the less popular insects such as cockroaches and mosquitoes surviving and thriving. Acknowledging that humanity's own actions are reducing acceptable animals to provide conduits to nature may have the potential to trigger an existential crisis and need for action. However, for meaningful change to be enacted, there must also be recognition that spirituality does not belong to humans alone. Separating spirituality from religion allows focus to shift to the experiential aspects of the concept that many animals could be capable of experiencing in their day-to-day lives if given the opportunity.

The examples provided in this section largely relate to mammals, those animals that humans perceive as phylogenetically closer to them. Reptiles, fish and invertebrates have received less consideration in the areas of both emotional capacity and spiritual possibilities, except when grouped under the generic concept of the natural world in its entirety. The next chapter examines spirituality from a cultural perspective where reptiles, fish and invertebrates do merit spiritual significance in some human cultures.

REFERENCES

Bekoff, M. (2001). The evolution of animal play, emotions, and social morality: On science, theology, spirituality, personhood, and love. *Zygon, 36*(4), 615–655. https://doi.org/10.1111/0591-2385.00388

Brooks Pribac, T. (2017). Spiritual animal: A journey into the unspeakable. *Journal of the Study of Religion, Nature and Culture, 11*(3):340–360. https://doi.org/10.1558/jsrnc.31519

Challenger, M. (2021). *How to be animal. A new history of what it means to be human.* Canongate Books Ltd.

Cunningham, P. (2022). The case for animal spirituality-Part 1: Conceptual challenges, methodological considerations, and the question of animal consciousness. *Journal for the Study of Religion, Nature and Culture, 16*(2), 186–224. https://doi.org/10.1558/jsrnc.18801

EFSA AHAW Panel (EFSA Panel on Animal Health and Welfare), Nielsen, S.S.,Alvarez, J., Bicout, D.J., Calistri, P., Canali, E., Drewe, J.A., Garin-Bastuji, B., Gonzales Rojas, J.L., Gortazar Schmidt, C., Herskin, M., Miranda Chueca, M.A., Padalino, B., Pasquali, P., Roberts, H.C., Spoolder, H., Stahl, K., Velarde, A., Viltrop, A., Winckler, et al. (2021). Scientific opinion on the welfare of sheep and goats at slaughter. *EFSA Journal, 19*(11), 6882.

Faver, C.A. (2009). Seeking our place in the web of life: Animals and human spirituality. *Journal of Religion & Spirituality in Social Work: Social Thought, 28*, 362–378. https://doi.org/10.1080/15426430903263161

Fisher, C.L. (2005). Animals, humans and x-men: Human uniqueness and the meaning of personhood. *Theology and Science, 3*(3), 291–314. https://doi.org/10.1080/14746700500317289

Ghaderi, A., Tabatabaei, S.M., Nedjat, S., Javadi, M., & Larijani, B. (2018). Explanatory definition of the concept of spiritual health: A qualitative study in Iran. *Journal of Medical Ethics and History of Medicine, 11*, 3.

Goodall, J. & Berman, P. (1999). *Reason for hope: A spiritual journey.* Grand Central Publishing.

Huber, J.T. & MacDonald, D.A. (2012). An investigation of the relations between altruism, empathy, and spirituality. *Journal of Humanistic Psychology, 52*(2), 206–221. https://doi.org/10.1177/0022167811399442

Hubert, H. & Mauss, M. (1981). *Sacrifice: Its nature and functions* (Revised ed.). University of Chicago Press.

Puchalski, C.M., Vitillo, R., Hull, S.K., & Reller, N. (2014). Improving the spiritual dimension of whole person care: Reaching national and international consensus. *Journal of Palliative Medicine, 17*(6), 642–656. https://doi.org/10.1089/jpm.2014.9427

Sen Nag, O. (2020, July 13). Rangila, the dancing bear's happy story is a landmark victory for animal rights. *Environment, World Atlas.* https://www.worldatlas.com/news/rangila-the-dancing-bear-s-happy-story-is-a-landmark-victory-for-animal-rights.html

Sheldrake, P. (2014). *Spirituality: A guide for the perplexed.* Bloomsbury Academic. https://doi.org/10.5040/9781472594532.ch-001

Taylor, B. (2005). *The encyclopedia of religion and nature.* Thoemmes Continuum.

Tobias, M. (2000). A gentle heart. In M. Bekoff (Ed.), *The smile of a dolphin: Remarkable accounts of animal emotions* (pp. 171–73). Random House/Discovery Books.

Willett, C. (2014). *Interspecies ethics.* Columbia University Press.

9

SPIRITUALITY AND CULTURE

While describing nonhuman animals as spiritual beings has the potential to offend and encroach on human-exclusive domains, there are some traditional cultural beliefs that imbue certain animals with spiritual significance. This chapter delves into some of those cultures and the impact this can have on an animal's life. It is by no means an exhaustive interrogation of cultural beliefs but does provide further context for the perception of animals as spiritual beings, often underpinned by their perceived emotional capacity. But first, it should never be forgotten that animals exist within their own cultural framework independent of any human culture.

Within their animal culture, observation or interaction with the group allows knowledge to be passed from parent to offspring, older to younger or younger to older, or between peers of the same generation in the form of an inheritance (Safina, 2020). Animal culture, according to Frans de Waal (2001), is a way "shared by the members of one group but not necessarily with the members of other groups of the same species" (p. 30). There can be systematic variations in knowledge, habits and skills between groups, including tendencies and preferences, attributable to exposure to and learning from others, rather than genetic or ecological causes. Social learning plays a key role in culture, enabling the rapid transmission of information through groups and giving rise to local variants that become embedded over time and generations. This means one group may behave very differently from another group, just as human cultural groups can be markedly different.

Positioning culture as a form of inheritance raises an important question to ponder. Assuming animals have their own culture in which emotional and spiritual capacities are enacted in a manner relevant to their unique needs, what is the long-term impact when deprived of social interactions to pass on these cultural norms? With the escalating forced dependence of individual animals, groups and species on humanity through commodification, isolation, habitat encroachment and environmental degradation, the lives of many animals are no longer balanced within the broader ecological scheme. Opportunities to embed experiences that increase meaning, purpose and promote flourishing—fundamental to animal spirituality—are diminished or extinguished from their daily lives.

Animal culture is not limited to mammals and birds in whom it may be more easily recognised. Reptiles, amphibians and invertebrates demonstrate their own unique cultures stored in the minds and actions of individual members and passed down

DOI: 10.1201/9781003298489-10

through generations. Crocodiles have survived in much the same format as their kin from 100 million years ago. Once dismissed as primitive and small-brained animals driven by programmed behaviours, it is now evident that they relate to each other according to their learned place in the dominance hierarchy, a ranking that can be flexible and allow for cooperation if required. They have been known to form semi-circles at the entrance of shallow pools on flood plains, creating a net of mouths for any fish that enter. The need to leave no gaps in the net over-rides any desire to fight (Benyus, 2014).

Insects may also show a more subtle form of culture, overlooked within the rigidity of their instinctual hard-wired behaviours and for some, short lifespan. This may be evident in their communication, methods of surviving and reproducing or more unique examples of agricultural and animal husbandry skills, paper making and net making. There are termites who grow fungus for food, ants who keep aphids as livestock, wasps who make paper from cellulose and caddisfly larvae who catch food in net-like webs (Sverdrup-Thygeson, 2019). Insects also have their standards, as evidenced by honeybees during the Australian heatwave of 2019. In the bee world, drunkenness is not tolerated, and when the hot weather fermented the nectar in some Australian flowers, drunk foragers were turned away from the hives until they sobered up (Millman, 2022).

Exploring animal spirituality within human cultures runs the risk of lapsing into anthropocentrism, especially when remembering that humans are animals and therefore should not hold a monopoly on culture. However, such is humanity's dominance over animals and the planet, it is necessary to give some thought to the historical context and positive or negative flow-on effects for animals in certain human cultures. Even where one human culture values the spiritual significance of certain animals, this can be subjugated to the needs of a more dominant human culture, as happened in Australia's Great Emu War of 1932.

CASE STUDY – THE GREAT EMU WAR OF 1932

A bird that grunts rather than tweets or whistles, runs rather than flies, stands almost 2 m tall, puffs up and hisses when threatened and appears on the Australian Coat of Arms sounds make-believe. But emus, a member of the group of flightless birds called ratites, range all through mainland Australia, from open plains to snowfields to forests and savannah woodlands. They rarely go near densely populated areas, rainforests or deserts, but the provision of water sources for the domestic stock has seen them increase in numbers in more arid areas. These nomadic birds, who can run up to 50 km/hour, also specialise in paternal care (Bush Heritage Australia, 2023). After the mother lays her large dark green eggs, she wanders off leaving the father to incubate the eggs for almost eight weeks with no time in his caring duties to eat or drink adequately. Father emus can lose kilograms in weight as they patiently await the arrival of their chicks. Fathers then stay with the young for two years, protecting them and teaching them all about emu culture—from how to find food to the best water holes and creeks to flop into and soak their feathers.

Despite their stature and speed, emus face threats such as habitat loss and destruction, fences which interfere with their movement and migration, and deliberate slaughter by humans. The damage they inflicted on wheat crops saw them labelled as destructive pests and killed in large numbers by European settlers to Australia. Emus were quickly eradicated from the island of Tasmania, but this proved more difficult in main-land Australia, especially in the wheat belt of Western Australia where mobs of emus would migrate to breed and flatten crops in their wake. By 1922, the West Australian government declared the emu as vermin status and offered a bounty for emu beaks. A decade later, the Great Emu War erupted, with troops from the Royal Australian Artillery deployed with machine guns and thousands of rounds of artillery to decimate the 20,000 marauding emus. However, the emus proved to be evasive and cunning adversaries, with no weapons vehicle able to keep up with their running skills (Garner Gore, 2016).

In sharp contrast, Aboriginal and Torres Strait Islander peoples perceive the emu as a special kind of person. To the Bininj people of western Arnhem Land in the Northern Territory, the emu is referred to as woman and contributes to their spiritual connection to animals and land (Garde, 2017). Emus are special people who wander the bush in search of food, always returning to places in the right season to get what they want. Importantly, the Bininji talk about emus from a unified perspective, com-prising a combination of biological, environmental, personal, experiential and religious domains. Emu is not an Aboriginal word, instead they are referred to by names such as Ngurrurdu, Kurdukadji, Alwanjdjuk, Durrk and Wurrbbarn depending on the Aborig-inal and Torres Strait Islander peoples' clan (Telfer & Garde, 2017).

The many different meanings and complex identities of the Australian emu are human constructs for a bird seeking to adapt to a shrinking habitat and enculturate the next generation into meaningful activities such as foraging for food, acting as seed dispersers in the natural ecosystem and soaking in the best waterholes. As vermin who trample crops or functional farmed commodities for medicine, oil and food, they are de-animalised to allow slaughter. As special people from when the world began, they maintain their spirituality through meaningful co-existence with indigenous cultures.

When viewed within the Zoological Emotional Scale, the emu's spirituality and cultural significance bestowed on them by the indigenous peoples was not sufficient to counter the dominant culture's perceptions of emus as foe or functional. The irony of attempting to massacre a bird who now adorns the Australian Coat of Arms illustrates the uneasy tension between humans and animals when the animal threatens human economic needs. It also highlights the imbalances between human cultures where an animal's spiritual significance is determined by the dominant cultural group rather than by the traditional group, the Aboriginal and Torres Strait Islander peoples, thus adding an additional layer to the racism experienced by this minority group. Figure 9.1 presents some of the competing positions this giant bird holds in contemporary Australian soci-ety, influenced by first-hand experience and cultural factors, as well as learned attitudes and illogical attributions of blame.

CULTURAL PERCEPTIONS OF ANIMAL SPIRITUALITY

The history of the emu provides some insight into how human perceptions can radi-cally alter across time and culture. Archaeological records show animals in art, architec-ture, folklore and the religious and spiritual rites of many ancient traditions. Animal symbols and metaphors have appeared in religious practices from different historical eras and geographical locations, and animal guardian spirits were often called upon to mediate healing and wellbeing. Animals were also sacrificed to give thanks or to

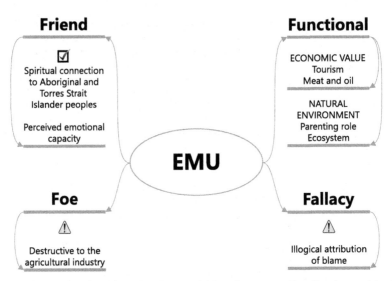

Figure 9.1 The Zoological Emotional Scale is applied to the emu to identify where spirituality and emotional capacity are attributed to this unique bird.

protect against evils. There are still some societies and cultural groups that value and live side by side with animals, imbuing animals with a spiritual presence to maintain balance in the world (Challenger, 2021). The peoples of the Arctic offer an example of humans and animals sharing personhood, with animals such as caribou and reindeer being regarded as "sentient beings able to hear, judge and react to human thoughts" (Anderson, 2004, p. 15).

Animism is an ancient form of human beliefs that perceives aspects of the natural world as containing living representations of the sacred (DeMello, 2012). Certain places, objects, plants and animals are infused with qualities of the divine and treated with particular reverence. Some cultures perceived animals as potential clan members, ancestors or intermediaries between the sacred and the irreverent worlds. In these cultures, humans did not attempt to control animals, rather to maintain proper relations with them. This did not negate the need to hunt them, but in some cases saw the hunter seeking the permission and pardon from the animal's spirit to kill them for food. In Australia, Aboriginal and Torres Strait Islander Elder Graham Paulson has been quoted as saying that in an animistic world, everything is interconnected—people, plants and animals—as part of a larger reality and a sharing of spirit. Humans are on equal footing with nature and are morally obligated to treat animals, plants and landforms with respect (Glynn-McDonald & Sinclair, 2021). Some animals are perceived as spiritual reflections and therefore as sacred as a relative.

Arising from animism, animal worship (zoolatry) developed into an independent form of worship that ranged from the reverence of animals as holy, to the physical worship of specific animals, to animal sacrifice. Aspects of zoolatry remain in place in some contemporary societies, although often competing with the needs and demands of more dominant social groups which would remove animals from shared spaces to allow humans with their modern advances to hold precedence. Animal worship did not automatically instil the creature with spirituality, rather it often focused on what the animal symbolised including its associations with either a specific deity or nature itself. Animals could be perceived as guides or totems, becoming spirits in the form of animals that would guide a human throughout life. Alternatively, in some cultures, animals were treated with respect because they could potentially contain the soul of a human being who had been reincarnated—or reborn—into an animal in their next life.

Humans have long relied on animals to provide a connection to nature but this has become increasingly difficult in contemporary times. Historian Richard Bulleitt (2005) wrote that the spiritual link previously existing between animals and humans has been destroyed through industrialisation, consumer capitalism and the increasing commodification of animals in many countries. This has led to contradictory feelings and actions towards animals. In some African cultures, elephants were revered as creatures that incarnated the human virtues of intelligence, wisdom and physical strength (Wood, 2020). From the Kamba tribe of Kenya to Gabon in West Africa to Ghana and Sierra Leone to Namibia, elephant wisdom was viewed as sacred, featuring heavily in cultural symbolism, art and storytelling, and elephants variously revered, feared or worshipped. This changed in later years when elephants became prized by poachers for their tusks or killed when they encroached on lands that humans had claimed for agriculture or development.

The tiger in Sumatra faced similar difficulties when killing livestock and people who had taken over their territory. However, among some communities, tigers continued to embody a range of spiritual attributes that transcended their animal form and provided a level of protection (McKay et al., 2018). This belief in spirit tigers was supported by oral traditions confirming the spiritual capacity of the animal and passed down through generations, reinforcing a strong relationship between human and tiger, or the ancestors that the tiger was thought to embody. Beliefs and opinions towards carnivores, however, can differ within and between human societies, cultures and historical times and may be negated for a number of reasons. Fear or economic losses arising from conflict over shared land, perceptions of monetary gain from the animal or animal parts or other perceived benefits such as sport and hunting can all contribute to their forfeit of spiritual status and their destruction in a human-centric world.

A common theme among these diverse traditions is inclusion of the natural world as a source of spiritual lessons, in sharp contrast to other worldviews that perceive nature as something to be dominated, subjugated and controlled (Faver, 2009). As more and more humans live in urban areas and leave behind the traditions of the past, a de-identified piece of packaged meat on a supermarket shelf has replaced knowledge and respect for the animal in its entirety.

COGNITIVE ETHOLOGY

When writing a foreword for the book *Navajo and the Animal People: Native American Traditional Ecological Knowledge and Ethnozoology*, William B. Tsosie Jr., Navajo traditionalist and anthropologist, provided a poignant description of how the perspective of the Navajos' connection with nature continues to shift under the pressure of living in a modern world. He describes the implications of losing the spiritual connection to the animals from which people emerged:

The natural world people like the Ch'osh, Na'ashoii, Tsidii, and the Naldlooshii—the insect, reptilian, bird, and animal people—have long been absent from the Navajo conscience. These people of the natural world and the Five-fingered People—humans—had a long shared relationship in the past. They lived together, interacting with and needing each other, looking out for and helping each other [...] Without a connection to the natural world, we as humans are disconnected and not complete. In modern times we have emptied our souls by not having a connection to the natural world.

(Pavlik & Tsosie, 2014)

Co-author Steve Pavlik believed that Native Americans had always been cognitive ethologists in that they recognised animals as conscious beings, bestowed with a wide range of emotions. Indigenous cognitive ecology was embedded within the sacred creation stories from the spiritual realm, thus making it more inclusive and offering a deeper, more comprehensive understanding of animal behaviour than Western science. Similar to the Australian Aboriginal and Torres Islander peoples who include reptiles and crustaceans such as the saltwater crocodile, crabs, turtles and yams in their deep spiritual connection with land, Native Americans also included the reptiles and insects that are so often omitted in general discussions of animal spirituality.

CONTINUUM OF ANIMAL SPIRITUALITY IN RELIGION

Endowing animals with spiritual meaning and significance is approached differently in the beliefs, writings and actions of many formal religions. Judaism and Christianity differentiate humans from other animals and bestow human domination over animals (Magliocco, 2018). This allows a division between body and soul where only humans are perceived as being gifted with souls. Islam positions humans as guardians of living beings while Buddhism advises cruelty of any kind to animals should be avoided (DeMello, 2012). Buddhist teachings do not differentiate humans from animals and apply the same moral obligations to each. Hinduism affirms that animals and humans both have souls, but humans remain superior to animals.

While there remains a perception that many of the larger non-Western cultures see both animals and humans as having access to a spiritual world, this is not always apparent as attitudes and needs change. A belief in the sacred and spiritual within the domesticated cow provides an example of the continuum within which one species may be positioned and how this can change in an increasingly interconnected world when traditions, values and demands blur. Some people who identify as Hindu perceive cows as sacred, worthy of reverence, and therefore to be cared for and protected (Valpey, 2020). However, India, where 80% of the population are Hindu but only 20% are vegetarian, ranked fifth among nations with the greatest number of cows slaughtered, suggesting that the legacy of India's special regard for the sacred meaning of cows may be changing (Challenger, 2021). The United Nations Food and Agriculture Organisation (2023) estimates over 300 million cows are slaughtered each year to meet human demand for meat, leather and other by-products, equating to almost 900,000 cows per day. Meanwhile, on Norfolk Island, a tiny island in the South Pacific Ocean, cows provide an interesting tourist feature as they enjoy the permission granted them in the 1800s to wander freely with right of way over traffic.

Jainism, a small Indian religious tradition, specifies that all souls are sacred regardless of species and extends the concept of nonviolence to animals, microbes, plants, micro-organisms and things with the potential for life (Rankin, 2020). The Jain tradition, based on instinctive understanding and a spiritual practice that preceded scientific findings of emotional capacity in animals, affirms the significance of all beings, irrespective of size or role, to the overall survival of life on earth. An important concept in Jainism is *Jiva*, defined as a unit of life and soul that is sentient and has the capacity for spiritual growth. *Jiva Daya* is a sympathy with all sentient life, allowing animal welfare to achieve an equivalent status to human welfare in Jain morality.

Perspectives that position animals as being created separately to humanity are also plentiful. The Christian Old Testament Bible, Genesis 1:26–27, clearly states that God created "man" in his own image and gave man dominion over fish, birds, animals—in fact, every creeping thing, thus reaffirming humanity's dominant position over mortal beings and uniqueness in possessing an immortal soul. A quote from C.G. Scruggs, former editor of *Progressive Farmers*, sums up the agricultural community's view on insects and their eradication: "Nature does exist for the convenience of man, for the Bible tells us so. That's why you will win!" (Lockwood, 1987, p. 74). In contrast, St Francis of Assisi, Christian patron saint of animals and the environment, argued for compassion and mercy for animals through stewardship rather than dominion as he reportedly preached to both humans and animals in his travels.

Once a year since 1985 an Episcopal Church in New York City opens its doors to greet hundreds, sometimes thousands in pre-COVID-19 years, of humans and animals. Animals of all shapes, sizes and behaviours line up in their carriers, boxes and on leashes to join the annual Blessing of the Animals service. This celebration of St Francis of Assisi and the Christian God's creatures is replicated globally in Christian churches. Most are not quiet events as animal emotion spills over in expressions of joy, exuberance, uncertainty, bravado or fear in the milling crowd of dogs, cats, birds, rodents, reptiles, farm animals and humans awaiting their blessing. Some churches provide affirmation of the animal's blessing in the form of the St Francis of Assisi charm shown in Figure 9.2. These services, intended to bring humans and animals together in a reflective moment of shared connection and increased awareness of animal rights, have attracted mixed commentary over the years. Speaking at the Episcopal Cathedral of St. John the Divine in New York City, Dr Hobgood-Oster, Professor of Religion and Environmental Studies, questioned whether these larger events were removed from reality, adding:

> One might question whether keeping animals in a large room full of the smells of incense, shouts of human voices, and presence of thousands of other creatures for over two hours is a good experience for the animals. Then again, how would a human being ever really make that determination?
>
> (Hobgood-Oster, 2008, p. 117)

RAINBOW BRIDGE

The contradictions evident in awarding some but not all animals a spiritual status within contemporary Western societies are evident in the growing popularity of the Rainbow Bridge trope that recognises an afterlife for those animals loved by humans (Magliocco, 2018). Rainbow Bridge, an imaginary bridge connected to the Christian

Figure 9.2 Some churches provide affirmation of the animal's blessing in the form of a St Francis of Assisi charm. Photograph supplied by the author.

Heaven, has emerged in popular culture as the place where beloved companion animals go after death to await the arrival of their human. Taken literally, this suggests that Rainbow Bridge may only be open to those animals living in close, familial relationships but who, according to Christian doctrine, were forbidden an afterlife and therefore could not enter Heaven alone.

Professor of anthropology and religion, Sabina Maggliocco, describes the legend of Rainbow Bridge as a meadow located "this side of Heaven", before the entrance to Paradise and connected to it by a rainbow. The dead companion animal, restored in body and health, plays peacefully in the sunshine with plentiful food and water as they await their human. Once their human dies, the companion animal senses this and bounds to the Rainbow Bridge for a joyous reunion. While the origins of this story remain unclear (sometimes attributed to a poem by Paul C. Dahm, other times to a book by William N. Britten, intended to comfort children at the loss of a companion animal), it has now become a popular image to buffer adult grief. After euthanising a beloved companion animal, many veterinary clinics routinely send a condolence card featuring the Rainbow Bridge image to the grieving human.

In many countries, companion animal ownership has increased with animals treated as kin with all the perks of personhood in life and death. Humans enter the relationship knowing they will most likely outlive more than one animal and experience a grief that some describe as of equivalence to grief for a human. Rainbow Bridge and other aspects of the growing pet mortuary business, including cremation, urns, burial plots and animal loss sympathy cards, offer solace and some semblance of control over the inevitability of death for those animals who have been granted human-like status.

However, many companion animals do not die in loving homes, instead treated as commodities to be discarded when inconvenient or no longer of value. Without human ownership to bestow meaning upon these animals—including entry into the mythical Rainbow Bridge—their lives and deaths can remain unnoticed and without spiritual significance. Perhaps, the true message of Rainbow Bridge lies not in the comfort it offers, but as a timely reminder that humanity's pervasive power to privilege certain animals over others extends beyond the living.

IMPLICATIONS

Attribution of both emotion and spirituality to animals risks sabotaging domains jealously guarded by humans, thus triggering cognitive dissonance where anomalies in treatment are identified. Where only the familiar or the beloved animals are awarded spiritual significance, or an animal's traditional spiritual significance is negated for economic or other discriminatory reasons, the resultant moral and ethical dilemmas can further compound the psychological discomfort of cognitive dissonance. History shows that humans frequently and unjustly change the ground rules, turning some animals of spiritual significance into pests, or allowing others to retain their spiritual significance on human terms. This one-sided relationship detracts from the animals' rights and access to activities that could allow them to flourish within their own culture as autonomous spiritual beings. It also discriminates against minority cultures whose perception and treatment of animals as spiritual beings may be overruled by other cultures. Dominant human groups across history have wielded the power to control both minority groups and their animals of spiritual significance, thus adding a nuanced layer of racism when traditional icons are thoughtlessly destroyed for economic reasons.

Untangling the hierarchy of human cultures that impact animal treatment can prove a complex task that is becoming increasingly urgent as natural environments degrade and human connection to animals and land erodes. Whether a person recognises the spiritual and emotional capacity of animals or not, there is an urgent need to re-assess the ethical treatment of animals as the world struggles with loss of biodiversity and changes in weather patterns. This chapter has provided only a snapshot of the spiritual role that animals hold in different human cultures. However, the examples provided suggest animals and nature are commonly acknowledged across cultures, religions and faiths, positioning a range of spiritual leaders and group members as natural advocates for animal welfare and conservation. Motivation to change can come from many different sources and for many different reasons, but recognising the need for change is crucial. Spiritual leaders may thus be in a strong position to move people to the action stage of change and consolidate a unified approach to animal wellbeing. This requires disputing the hierarchical positioning of human economic requirements over animals and traditional human cultures, interrogating and building on tokenistic concepts such as Blessing of the Animals and Rainbow Bridge and valuing animals and nature for themselves rather than what they can offer humanity.

REFERENCES

Anderson, D. (2004). Reindeer, caribou and 'fairy stories' of state power. In D. Anderson & M. Nuttall (Eds.), *Cultivating Arctic landscapes: Knowing and managing animals in the circumpolar North* (pp. 1–16). Berghahn.

Benyus, J.M. (2014). *The secret language of animals.* Black Dog & Leventhal Publishers, Inc.

Bulleitt, R. (2005). *Hunters, herders, and hamburgers: The past and future of human animal relationships.* Columbia University Press.

Bush Heritage Australia (2023). *Emu.* Emus - Bush Heritage Australia

Challenger, M. (2021). *How to be animal. A new history of what it means to be human.* Canongate Books Ltd.

DeMello, M. (2012). *Animals and society: An introduction to human-animal studies.* Columbia University Press.

de Waal, F. (2001). *The ape and the sushi master.* Allen Lane - Penguin Books.

Faver, C.A. (2009). Seeking our place in the web of life: Animals and human spirituality. *Journal of Religion & Spirituality in Social Work: Social Thought, 28,* 362–378. https://doi.org/10.1080/15426430903263161

Garde, M. (2017). Introduction. In M. Garde (Ed.), *Something about emus. Binjini stories from Western Arnhem Land* (pp. xxv–xxiii). Aboriginal Studies Press.

Garner Gore, J. (2016, October 18). Looking back: Australia's emu wars. *Australian Geographic.* https://www.australiangeographic.com.au/topics/wildlife/2016/10/on-this-day-the-emu-wars-begin/

Glynn-McDonald, R. & Sinclair, R. (2021). Connection to animals and country. *Common Ground.* https://www.commonground.org.au/learn/connection-to-animals-and-country

Hobgood-Oster, L. (2008). *Holy dogs and asses: Animals in the Christian tradition.* University of Illinois Press.

Lockwood, J.A. (1987). The moral standing of insects and the ethics of extinction. *The Florida Entomologist, 70*(1), 70–89. https://www.jstor.org/stable/3495093

Magliocco, S. (2018). Beyond the Rainbow Bridge: Vernacular ontologies of animal afterlives. *Journal of Folklore Research,* 55 (2), 39–68. https://doi.org/10.2979/jfolkrese.55.2.03

McKay, J.E., St. John, F.A.V., Harihar, A., Martyr, D., Leader-Williams, N., Milliyanawati, B., Agustin, I., Anggriawan, Y., Karlina, Kartika, E., Mangunjaya, F., Struebig, M.J., & Linkie, M. (2018) Tolerating tigers: Gaining local and spiritual perspectives on human–tiger interactions in Sumatra through rural community interviews. *PLoS ONE, 13*(11), e0201447. https://doi.org/10.1371/journal.pone.0201447

Millman, O. (2022). *The insect crisis: The fall of the tiny empires that run the world.* W.W. Norton & Company, Inc.

Pavlik, S., & Tsosie, W. B. (2014). *Navajo and the animal people: Native American traditional ecological knowledge and ethnozoology.* Fulcrum Publishing.

Rankin, A. (2020). *Jainism and environmental politics.* Routledge.

Safina, C. (2020). *Becoming wild. How animals learn to be animals.* Oneworld Publications.

Sverdrup-Thygeson, A. (2019). *Extraordinary insects.* Mudlark.

Telfer, W. & Garde, M. (2017). Appendix 1. In M. Garde (Ed.), *Something about emus. Binjini stories from Western Arnhem Land* (pp. 147–152). Aboriginal Studies Press.

United Nations Food and Agriculture Organisation. (2023). *Crops and Livestock Products.* FAOSTAT. https://www.fao.org/faostat/en/#data/QCL

Valpey, K.R. (2020). *Cow care in Hindu animal ethics.* Springer Nature. https://doi.org/10.1007/978-3-030-28408-4_1

Wood, L. (2020). *The last giants.* Hodder & Stroughton.

10

ANIMAL MORALITY AND MORAL STANDING

Contemplating the possibility that animals have their own unique emotional and spiritual capacity raises questions about the flow-on effect on their moral standing. Moral standing, also known as moral status, is held by any entity whose continued existence and wellbeing are ethically desirable (Schönfeld, 1992). Possessing moral standing can also determine if and to what degree that entity receives concern and respect when decisions are being made about them. Humans make decisions that affect animals directly or indirectly in a range of contexts, largely from a position of perceived power over that animal and the environment. High moral standing promotes inclusion of the animal's perspective when wielding that power. Minimal or no moral standing relegates the animal's perspective to inconsequential and can perpetuate suffering. The following tale of a turtle describes the indirect but detrimental impact on an animal's moral standing and wellbeing through humanity's control over even the remotest of environments.

THE TALE OF A TURTLE

Sea turtles have contributed to the health of the oceans for over 100 million years. Described as a keystone species, every part of their life cycle contributes something to the ecosystem, from providing nutrients to coastal vegetation from leftover eggs or dead hatchlings, to keeping seagrass healthy and managing jellyfish numbers. Some of the seven species are also being pushed to critical endangerment and extinction through their vulnerability to human activity (International Fund for Animal Welfare, 2022). Sea pollution, entanglement in fishing gear, warming of waters and destruction or degradation of feeding and breeding grounds have combined with human demand for sea turtle meat, leather, shell products and illegally traded exotic pets to reduce the number of hatchlings. This, in turn, impacts natural predators and the coastal and marine vegetation. Six of the seven species of sea turtles are found in Australia, where coastal Aboriginal and Torres Strait Islander peoples hold strong cultural, social and spiritual ties to these important creatures. European settlers commercially harvested sea turtles for meat, eggs and shells between the mid-1800s and mid-1900s, resulting in a significant decline in numbers. The Commonwealth of Australia's (2017) *Recovery Plan for Marine Turtles in Australia* offered some protection for these culturally and environmentally significant animals, but their numbers continued to decline.

DOI: 10.1201/9781003298489-11

Sea turtles tend to be solitary reptiles with the typical absence of expressive facial mobility that humans find so endearing in mammals. Along with many other reptiles, turtles can be overlooked in discussions of emotional capacity due to lack of a familiar point of reference and anthropocentric expectations of what emotions should look like. In 2022, a tiny green sea turtle hatchling missing one flipper was found lying on his back in a rock pool on an Australian beach. Weighing only 127 g, with a chip in one of his three remaining flippers and a hole in his shell, this tiny hatchling achieved international fame for a very strange reason. After being sent to Sydney's Taronga Zoo for rehabilitation, the little turtle defecated plastic in various colours, textures and sizes for the next six days (Sonnenschein, 2022). While his other injuries could be attributed to natural predators, the potentially fatal ingestion of so much plastic as "baby's first food" highlighted human culpability in the mass extinction of animal species. The tiny turtle was not unique in pooping plastic. Four years earlier on the other side of the Australian continent, a one-year-old Loggerhead sea turtle gained social media fame when he too was pictured beside a pile of defecated plastics and tangled wires after being rescued and rehabilitated at Perth Zoo.

While headlines and social media focused on the sensationalism of pooping plastic, the Taronga Zoo hatchling also became the haunting face of a pervasive human-driven disaster that had been decades in the making. The tiny turtle was an innocent victim whose ingestion of potentially fatal human plastics shocked the world into acknowledging that this damaged little turtle should have moral standing because he and his ocean friends had a moral right to life for their own sakes. He was also a visual reminder of an epoch labelled the Anthropocene.

ANTHROPOCENE

The twenty-first century has seen this word increasingly enter conversations of climate change and human accountability after a 2002 article in *Nature* pointed a finger directly at humanity's escalating effects on the environment. Dutch meteorologist, atmospheric chemist and Nobel Prize winner Paul Josef Crutzen wrote

> The Anthropocene could be said to have started in the latter part of the eighteenth century, when analyses of air trapped in polar ice showed the beginning of growing global concentrations of carbon dioxide and methane. This date also happens to coincide with James Watt's design of the steam engine in 1784.
>
> (p. 23)

Derived from the Greek terms *anthropo* meaning human, and *cene* meaning new, it describes an epoch in which humans have become the dominant force shaping the world and, therefore, animals and ecosystems on the planet. Positioning animals below humans is symptomatic of the Anthropocene, and reminiscent of Melanie Challenger's assertion that: "[…] we don't know the right way to behave towards life. This uncertainty exists in part because we can't decide how other life forms matter or even if they do" (p. 2).

Mounting evidence confirms humanity's potentially irreversible impact on the planet resulting in mass extinctions, loss of biodiversity, changing environments and ecosystems and millions of tons of plastics. So enthusiastically did humanity embrace plastics, their ubiquitous waste presence has been described as a key marker of the Anthropocene. It is estimated that around 75% of environments on land and 66% of marine environments have been significantly altered by human actions and around one million animals and plants are threatened with extinction (Pavid, 2019).

The tiny sea turtle's moral standing was embraced nationally and globally as symbolic of humanity's thoughtless and destructive wastage. Chunks of hard, sharp plastic piled beside his frail body invited the viewer to imagine the suffering imposed on this tiny sentient being who had barely started life. However, where human and animal lives intersect on a broader scale and their interests are in conflict, humanity's self-appointed exceptionalism has proved time and again to be sufficient moral justification to invalidate an animal's needs. This includes industrial animal farming, degradation or encroachment on natural habitats, confinement of animals and their commodification for entertainment or other human needs. These examples of human–animal interactions position animals below humans on an emotional, spiritual and moral level as to do otherwise risks cognitive dissonance in treating sentient creatures in these ways.

SPECIESISM

Responding to the Anthropocene is beyond the scope of this book, but it will entail challenging concepts such as speciesism. Speciesism is the practice of seeing one's species as morally more important than members of other species and finding ways to justify that perception. Human proponents of speciesism position nonhuman animals as inferior to humanity.

English philosopher Richard Ryder (2011) proposed two types of speciesism. Weak speciesism rationalised the exploitation of animals based on their perceived lack of morally important qualities, including high intelligence, reason, autonomy, a moral sense and a soul. Strong speciesism was less directive, simply validating the exploitation of animals because they were not human. Other philosophers and ethicists provided similarly themed explanations, with Lori Gruen (2021) contending that some humans do believe their humanity is distinct from—even transcending—their animality, a claim analogous to strong speciesism and human exceptionalism. Australian philosopher Peter Singer, who popularised the term speciesism in his 1977 book *Animal Liberation*, argued against the perception that being human was sufficient to have greater moral worth than animals. Instead, approaching this from a utilitarian perspective suggested that if animals were capable of pleasure and pain, they should have moral consideration irrespective of species.

Speciesism is endemic in human society and particularly evident where less moral standing is awarded to low-status animals such as food animals and disgust-provoking animals that fail to trigger a person's nurturing or emotional response. The resultant moral differentiation between a factory-farmed pig and a beloved companion dog, both of whom share similar mental and emotional capabilities, is paradoxical. The moral differentiation between a tuna fish who bears no physical resemblance to a mammal and lives in an alien underwater environment and a fluffy fledgling magpie sees one caught by the millions for human consumption and the other protected and nurtured by both parent bird and the human in whose garden the baby may inadvertently land.

Speciesism and anthropocentrism have the power to legitimise discriminatory treatment especially when tacitly validated by the hierarchical taxonomy of animal species. Protections based on this hierarchy may still have death as an end point but arriving at death can differ in levels of pain and suffering. Most domestic farmed animals have the right to be unconscious before their throat is slit and they are bled out, whereas tuna have started the painful and distressing suffocation process out of water before being killed. The preceding chapters have highlighted that many animals do possess consciousness, sentience and emotional capacity, even if behavioural manifestations of these attributes may differ from human expectations. Perceiving differences as indicators of absence may, in fact, be a form of speciesism demonstrating a person's lack of capacity or desire to delve into a world different to their own. This, in turn, removes any motivation to begin the process of change that requires transitioning from pre-contemplation, contemplation and preparation through to action and maintenance.

MORALITY

To acknowledge that animals may have emotional or spiritual capacity requires a shift in thinking from human benevolence bestowing or withholding an animal's moral standing to perceiving animals as moral subjects in their own right and capable of acting with moral motivations. Marc Bekoff (2007), a strong proponent of morality among animals, described the phenomenon as "a wide-ranging biological necessity for social living" (p. 87) and including basic components of cooperation, empathy, fairness, justice and trust. These basic components are displayed among vertebrate and invertebrate animals alike, revealing a capacity for moral actions guided by species-specific social norms for group living.

Researcher Oliver Curry (2016) suggested a definition of morality that appeared to have precursors in animal behaviour when he argued that "morality turns out to be a collection of biological and cultural solutions to the problems of cooperation and conflict recurrent in human social life" (p. 29). While limiting this definition to humans, he noted that rules related to kinship, mutualism, exchange and methods of conflict resolution appeared in nearly all societies and were also evident in animal behaviour. With animals capable of exhibiting all these rules to varying degrees and specific to their own animal culture, this definition is therefore equally valid for both humans and animals, a notion supported by philosopher Mark Rowlands (2012). Rowlands argued that animals can behave on the basis of moral motivations, including sympathy and compassion, kindness, tolerance and patience, and importantly "a sense of what is fair and what is not" (p. 32). If animals can act on the basis of moral motivations, this infers they are, in fact, moral subjects with moral standing, a title that humans have long considered a distinguishing feature between human animals and nonhuman animals.

Integrating these reflections suggests morality in its most basic form is a prosocial behaviour aimed at promoting (or at least not negatively affecting) the welfare of others. Embedded within social connections and stemming from a set of internalised rules, morality among humans and animals alike regulates interactions—especially those related to wellbeing or harm—within social groups and the social norms of what is right and what is wrong. It is therefore a dynamic social phenomenon among animal groups, just as it is among human societies. Bekoff raised the point that it was impossible for morality to suddenly appear in humans, rather it is an evolutionarily ancient, evolved trait fundamental to cohesion within a group.

Nevertheless, the debate over animal morality has a long and contentious history. Frans de Waal, Marc Bekoff and Jessica Pierce focused on a series of psychological

capacities as indicative of morality, including empathy, altruism, reciprocity, cooperation, sharing and fair play. Bekoff and Pierce (2009) grouped these into three clusters of
morality: the cooperation cluster (altruism, reciprocity, honesty and trust), the empathy
cluster (sympathy, compassion, grief and consolation) and the justice cluster (sharing,
equity, fair play and forgiveness). Lori Gruen (2021) felt that the ability to feel pain, to
suffer, to experience the moral emotions of love and concern, and the prosocial tendencies of cooperation, trust, reciprocity, conflict resolution and altruism, were fundamental to both animal and human morality. Gruen added that the opportunity to live
and play freely, enjoying experiences that make life feel valuable should not be exclu-
sive to human morality. Animals also have the capacity for these experiences when
given the opportunity free of human restraint. Synthesising these understandings suggests a moral being has the capacity to respond without irrational judgement and the
capacity to be good without needing to articulate what good means (Andrews, 2012).
It may also mean that animals are more moral than some humans whose irrational
judgements and attempts to allay cognitive dissonance may sabotage the fundamental
meaning of morality (Gruen, 2021).

Logically, this should award animals the same level of moral standing and inviolable
rights to life and liberty applicable to many (but not all) humans. It should also remove
the automatic positioning of animals as instruments for human wellbeing, whether
that be as food, sacrifice or comfort. A counter argument to this assumption wielded
by proponents of speciesism is that the granting of inviolable rights requires more than
consciousness, sentience or emotion. There should be some higher cognitive capacity
in order to see animals as equal to humans.

The UK *Animal Welfare (Sentience) Act 2022* that deemed invertebrates such as
crustaceans capable of suffering and therefore no longer to be boiled alive stopped
short of legislating that as sentient creatures, they should not be captured and killed
for human consumption. A common perception is that higher cognitive capacities
grant humans implicit ownership rights over beings who lack these capacities. This
species-based inequality is enshrined in the institutions of law, government and property (Bradshaw, 2021), not just in personal biases evident when assessing subjective
perceptions with instruments such as the Zoological Emotional Scale. Humanity's
longstanding tendency to treat animals as morally less than humans, thus legitimising
killing, confinement, hunting and habitat encroachment, may in fact be an implicit
strategy to avoid the cognitive dissonance of maintaining these practices in the face
of animal suffering.

The inclusion of both biological and cultural components to morality allows for
variations across human and animal cultures depending on social norms and learned
behaviour. As with animal emotion and spirituality, interpreting moral behaviours
through a human lens is flawed and disrespects their meaning within the context of
the animal's life experiences. The following behaviours of Cody, a Cypriot rescue dog,
illustrate the complexity of seeking to understand moral behaviours in different species, even the ones with whom humans are most familiar.

MURDER OR GRIEF?

Feathers drifted lazily in the afternoon breeze, sometimes coming to rest and stick in
the blood that covered the paving stones. Inside the house, remnants of a bloody carcass

stretched the length of the kitchen floor. Cody, a mixed-breed dog born in Cyprus and rehomed in the United Kingdom, sat quivering in a corner, seemingly shocked at her instinctual reversion to a life she had left behind six years ago. Delivered to a Cyprus veterinary clinic in a metal cage and abandoned when the puppies she was struggling to deliver were born dead, Cody found herself desexed, dewormed, rehabilitated and sent to the United Kingdom for rehoming.

Cody's life on the streets had been hard. Judging by her emaciated body, she had been hungry a lot of the time, probably forced to scavenge and kill to supplement the scraps provided by any passers-by who felt sorry for her. During rehabilitation, she showed no signs of connection with other dogs or the humans who were attempting to socialise her, refusing to make eye contact and seemingly unable to transition from an unwanted dog to a sentient being worthy of affection. After months of patient human contact, Cody was declared fit to be rehomed and she settled into domestic life with a small elderly terrier and a small elderly man. It was the terrier who completed Cody's enculturation into her new social group, soothing her fight-and-flight responses and the need to dominate at food time. As Cody's old terrier friend grew less mobile and slower to finish his meal, she never once tried to steal his food. The day after her terrier friend went to the veterinary clinic to be euthanised, the backyard massacre occurred.

By human standards, Cody committed an immoral act when she killed another creature for no apparent reason. It was many years since Cody, hungry and cold, had killed to satisfy her chronic hunger, making the deliberate attack on this bird seemingly inexplicable. Cody's behaviour, however, could be equally indicative of her confusion at the loss of her mentor. Cody's reliance on the old terrier was undeniable, and his loss temporarily removed structure and stability from her new life. Unable to articulate her emotions in human language, Cody's actions suggested an instinctual reversion to the old behaviours that had once sustained her. As Cody's shocked and grieving human narrated this story with Cody by his side, my mind questioned whether the bird's death expressed Cody's need to soothe the emotional pain of a grief she was ill-equipped to handle. While this may have been anthropomorphic conjecture, it was no less credible than assuming Cody lacked morality and remained on the verge of reverting to base instincts rather than the cooperation, empathy and justice she had learned to embrace in her new life (Figure 10.1).

Cody's tale is a timely reminder that it is only human arrogance that seeks to equate an animal's social, emotional and moral life with human models. Carl Safina learned this lesson while watching elephants with Cynthia Moss (2015). He admitted:

> I'd somehow assumed that my quest was to let the animals show how much they are like us. My task now—a much harder task, a much deeper task—would be to endeavour to see *who* animals simply *are*—like us or not.
>
> (p. 13)

Cody had experienced a life most humans could never imagine. That she had settled into a domestic life of security, stability and companionship was a testament to the resilience of animals rather than the ability of humans to reshape animal nature into their perception of morality. To understand Cody's behaviours and emotions required

Figure 10.1 Cody leans against the side of a chair as her elderly human rests his hand on her shoulder. Photograph by the author.

looking beyond the assumptions and constructs of human morality to a holistic perspective of morality as a species-specific prosocial behaviour.

MORALITY AND SOCIAL LIVING

Morality as a prosocial behaviour and an adaptive strategy for social living automatically positions this elusive concept as a natural component of animal culture. Where natural expressions of culture are impeded by human control, animals may lose the opportunity for the rich social lives inherent in their socially learned, collective ways of behaving. When analysing the moral rights of animals, Will Kymlicka and Sue Donaldson (2014) identified the human constraints imposed on three types of animals: domesticated animals, including companion animals and farmed animals who should be considered co-citizens; wild animals who live relatively independently as autonomous beings; and liminal animals, called denizens, including rats, possums, racoons, deer and many birds who are not domesticated but who must adjust to human activities to survive in the world. Domesticated animals differed from the other two categories through the human interventions that had made them dependent on human care. By necessity, they slotted into human schemes of cooperation, effectively removing any possibility of independent living. This resulted in a precarious existence whereby they depended on human beneficence and were unable to survive independently if and when relinquished intentionally or through unavoidable circumstances such as warfare and natural disasters. Kymlicka and Donaldson provided the much-publicised example of the precariousness of human domestication in Bill and Lou, two oxen who worked at Green Mountain College in Vermont (USA) as part of the college's sustainable agriculture programme.

THE STORY OF BILL AND LOU

Bill and Lou were two stoic oxen who had for 10 years ploughed the college field in tandem. When Lou hurt his leg on the job and could no longer work, there was a reported suggestion that both oxen be slaughtered and served as ox burgers in the cafeteria to demonstrate sustainable agriculture. The reasoning was twofold: once unable to work Lou was deemed useless and a drain on resources; and after 10 years together, Bill would miss Lou if they only killed Lou. The contradictions in this logic seemed self-evident to all but the College as they simultaneously positioned Bill and Lou as functional commodities to be discarded when serving no productive purpose and emotional beings who had formed an attachment bond that would cause grief when broken. To shift this illogical decision from the immoral to the moral, the College argued that not only would the unproductive and lonely Bill use resources, but he would also emit greenhouse gases with no resultant benefits to the humans. While the decision to euthanise Lou should not be questioned without knowledge of the extent of his injury, it was Bill's future that raised the ire of the global animal welfare community. Protestors posted images of the doomed Bill and Lou on social media, often snuggled up to a caring human who was trying to save their lives. Through these emotive images and the accompanying tragic story, Bill and Lou became individuals in need of rescue from human commodification under the guise of sustainability.

The contradictory positioning of Lou and Bill on the Zoological Emotional Scale in Figure 10.2 demonstrates the precarious existence of domesticated animals whose friend status can easily be over-turned when no longer functional. The rhetoric around Bill and Lou's future pitted sustainable agriculturalists against animal welfare lobbyists, and despite a nearby sanctuary offering both oxen a peaceful retirement home free of charge, Lou was subsequently euthanised in 2012. Lou and Bill were contributing members of a human community but they were still perceived as property, requiring other humans who did not own them to fight (unsuccessfully) for Lou and Bill's moral rights and control over life and death.

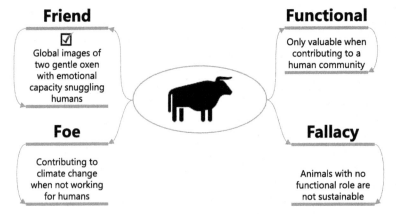

Figure 10.2 The Zoological Emotional Scale applied to Bill and Lou identifies the contradictory perspectives applied to these animals once they are no longer perceived as functional.

INVERTEBRATES AND MORAL STANDING

Many invertebrate animals also occupy a questionable space when examining their moral status and right to live. Arthropods, including insects, crustaceans, spiders and the many-legged millipedes and centipedes, provide a case of conditional moral consideration despite being the subject of increasing investigation into their sentience, cognitive capacities, neurological complexities and flexible behaviours. The moral exclusion of arthropods, reinforced by humanity's collective tendency towards fear verging on phobia for many insects and spiders, promotes behaviours contributing to the mass destruction of these essential creatures. Insects and spiders are commonly perceived to lack the cognitive, emotional and spiritual properties more readily attributed to mammals and birds, thus failing to achieve the moral standing of beings worthy of concern and respect. Bees are sometimes endowed with higher moral standing, being viewed as sociable and productive role models who provide a service to humanity. While this reduces their risk of intentional human-instigated immoral acts, they are still subject to harm through their value as a commodity of pollination.

In 1987, Jeffrey Lockwood, entomologist and philosopher, drew attention to the moral standing of insects with the statement that

> To deal responsibly, justly, and sensibly with issues of extinction and the moral status of insects requires a blending of ethics with biology. Fundamentally, we must establish a philosophically sound, scientifically consistent, ethical basis for our concerns regarding insects.
>
> (p. 71)

After analysing the philosophical basis of morality, Lockwood questioned the rationality of excluding animals from moral deliberations based entirely on differences between human animals and nonhuman animals. To differentiate required demonstration of valid and relevant moral differences, rather than the undefended criterion of species and speciesism. Despite this logic-based evaluation, Lockwood himself observed an overall aversion to acknowledging insects as beings deserving of moral consideration for themselves and proposed a minimum ethic based on their self-awareness, planning and pain: "We ought to refrain from actions which may be reasonably expected to kill or cause nontrivial pain in insects when avoiding these actions has no, or only trivial, costs to our own welfare" (p. 83), a timely reminder in the face of the mass insect extinctions associated with the Anthropocene.

A decade later, entomologist Michael Draney (1997) drew attention to the anthropocentric biases inherent in humans labelling an organism—especially insects—a pest when its net instrumental value to humans was negative. Supporting Lockwood's contention that insects are sentient and therefore merit a claim on moral consideration, Draney pointed out that applying moral rights to insects labelled as pests was negated by most ethical systems that assumed human interests and rights superseded the rights of insects. However, eradicating a pest population had deeper moral significance through the potential negative impact on the natural ecosystem, and ultimately flow-on effects to humanity and the planet.

In the 1970s, the deep ethology movement entered the philosophical community. This environmental philosophy positioned all living things as having an intrinsic or inherent value that was not dependent on serving some purpose for humanity (Glasser,

2011). The subsequent concept of deep reflective ethology invited humans to be more aware of what they did to nonhuman animals, and of their moral and ethical obligations to these creatures. Recognition that animals are sentient creatures with decisional and volitional capabilities and the capacity to experience positive and negative emotions and suffer pain is fundamental to the deep ethology movement and the antithesis of speciesism. This has redefined the meaning of animal welfare and triggered feelings of guilt for an increasing number of people when contemplating the immoral treatment of animals across cultures and times.

Many animals can and do act altruistically, with a capacity for empathy, trust and reciprocity. These behaviours are at the core of morality and allow animals to live cooperatively in their own societies. Nowhere is this more evident than in the farmed animal sanctuaries which are redefining the meaning of animal welfare, one animal at a time. Treating animals as moral beings with moral standing and worthy of respect is fundamental to animal sanctuaries such as the one that attempted to offer a home to Bill and Lou. It is a sad reflection of human society that animals must still be rescued from slaughterhouses and factory farming, but the existence of such sanctuaries is also a testament to changing attitudes and a willingness to improve the lives of animals one by one. Anecdotes from carers at these not-for-profit sanctuaries indicate that when freed from the terrors inflicted by uncaring humans, these rescued animals co-exist peacefully and often support each other to adapt to a new life of freedom. Rehabilitating at times traumatised new arrivals can require self-sacrifice and altruistic practices by both humans and animal residents until physical and emotional healing can begin in the safety of the sanctuary.

CASE STUDY–EDGAR'S MISSION, LANCEFIELD, VICTORIA, AUSTRALIA

Step through the entrance to Edgar's Mission sanctuary and you enter a magical world where being greeted by a wobbly goat in a psychedelic jacket feels completely natural. Overshadowed by the Macedon Ranges in Victoria, Edgar's Mission transports visitors to a safe land where every animal has a story, a personality and a new life (Figure 10.3). Two mini donkeys stroll down a leaf-strewn path as smiling visitors and hard-working volunteers step aside, and a Kelpie who never made it as a sheep dog mocks the concept of speciesism as she greets chickens, goats, sheep, cows and humans alike with a grin and a joyful sweep of her tail.

Edgar's Mission sanctuary, home to Clarabelle, Valentine, Lemonade, Gracie and Dottie who appeared in earlier chapters, is one of those special places where farmed animals are allowed to live, age and die peacefully and with dignity. No longer faceless, nameless commodities, each animal has the freedom to live a rich emotional life surrounded by family and friends and to leave a lasting impression when they die. Named in honour of the first resident, a pig called Edgar Alan Pig, Edgar's Mission has changed the lives of thousands of animals since 2003. It has also changed the attitudes of humans as they become engrossed via social media platforms or visits to the sanctuary in the daily lives of these unique individuals who have families, friends, diverse personalities and a zest for living irrespective of the traumas inflicted in their earlier lives.

Morality can be a hard concept to define, but prosocial behaviour aimed at promoting the welfare of others provides a starting point. This makes morality a social

Figure 10.3 The entrance to Edgar's Mission, Victoria, where a donkey grazes peacefully by the sign. Photograph taken by the author at Edgar's Mission Farm Sanctuary, Lancefield and reproduced with their permission.

phenomenon that regulates interactions between groups, especially those related to wellbeing and harm, and what is right and what is wrong. Under the banner "If we could live happy and healthy lives without harming others, why wouldn't we?", images and stories shared by Edgar's Mission leave the viewer in no doubt that animals have moral standing as they respond to each other without irrational judgement in the safety of the sanctuary. Demonstrations of care and empathy are not confined to an animal's own species, as contact and friendships between pig and sheep, dog and lamb, cat and lamb and more are poignantly captured in social media images that need no words.

IMPLICATIONS

Accepting that many animals—not just the ones that humans choose to love or identify with—are moral beings capable of demonstrating cooperation, empathy and justice has the anxiety-provoking potential of requiring people to re-think industrial food systems, reshape consumption patterns and re-consider the concept of animals as human property. Alternatively, people may disconnect so as to avoid the cognitive dissonance associated with simultaneously consuming animal products and loving nature. Those who act on this dissonance through life changes such as veganism may face what psychologist Clare Mann (2018) describes as "vystopia": the struggle to live in a non-vegan world with others who seem not to care and instead collude with the society's systematic animal abuse through their consumer choices. Many people who adhere to veganism have endured subtle or not-so-subtle complaints from family, friends and colleagues deriding the difficulties of catering for a vegan guest or mocking the perceived stringent requirements of this life choice. This forces the person to conduct an internal debate on how to counter these disrespectful comments without alienating family and friends, a battle that can rarely be resolved satisfactorily as strategies to neutralise the cognitive dissonance of meat eating are deeply embedded in society and aggressively protected.

However, there are changes underway, as the growing number of movements dedicated to animal welfare and rights demonstrate. These movements direct public attention to the multiple actions that disregard the moral standing of animals; for example, the use of gestation crates for pigs and calf crates (Humane Society of the United States), puppy farms (Lucy's Law and Oscar's Law), animal testing laboratories, the clothing trade and entertainment industry (People for the Ethical Treatment of Animals), the dog meat trade (Humane Society International), abuse in the dairy industry (Mercy for Animals) and the restraint of bears for bile farming (Animals Asia).

Empathy remains firmly embedded within the descriptor of morality and moral standing, supporting claims for animal welfare that rely on a shared capacity to recognise happiness versus suffering within and between species. For example, human empathic understanding of the meaning of pain coupled with scientific validation that lobsters could feel pain when boiled alive provided a catalyst for the legislation to change this human-inflicted suffering. However, animals perceived as foe may quickly lose moral standing when threatening human lifestyles or economies. An elephant, bear, wolf, emu, spider or termite who intrudes on human space loses moral standing and is at risk of indiscriminate death. To remain worthy of moral standing may require an animal abiding by its human-defined space and role, thus perpetuating implicit speciesism.

Awarding animals high moral standing is a necessary step if animals are to regain their rightful positions in a human-dominated world. Disregarding the needs and rights of animals, symptomatic of the Anthropocene, has already had far-reaching negative effects on the planet. Grassroots organisations have made a start, but individuals can play a key role through their day-to-day actions and choices. The following chapters explore some specific cases and their implications when animals are denied moral standing and ways that the Zoological Emotional Scale can help identify stereotypes, biases and gaps in animal wellbeing.

REFERENCES

Andrews, K. (2012). *Do apes read minds? Toward a new folk psychology*. MIT Press.

Bekoff, M. (2007). *The emotional lives of animals: A leading scientist explores animal joy, sorrow, and empathy—and why they matter*. New World Library.

Bekoff, M. & Pierce, J. (2009). *Wild justice: The moral lives of animals*. The University of Chicago Press.

Bradshaw, K. (2021). Human as animals - Pluralizing humans. *Utah Law Review, 1*, 185–209. https://doi.org/10.26054/0D-4TQQ-35C7

Challenger, M. (2021). *How to be animal. A new history of what it means to be human*. Canongate Books Ltd.

Commonwealth of Australia. (2017). *Recovery Plan for Marine Turtles in Australia 2017-2027*. https://www.nature.com/articles/415023a

Crutzen, P.J. (2002, January 3). Geology of mankind. *Nature, 415*(23). www.nature.com

Curry, O.S. (2016). Morality as cooperation: A problem-centred approach. In T. K. Shackelford & R. Hansen (Eds.), *The evolution of morality* (pp. 27–51). Springer. https://doi.org/10.1007/978-3-319-19671-8_2

Draney, M.L. (1997). Ethical obligations toward insect pests. *Ethics and the Environment, 2*(1), 5–23. https://www.jstor.org/stable/27766028

Glasser, H. (2011). Naess's Deep Ecology: Implications for the human prospect and challenges for the future. *An Interdisciplinary Journal of Philosophy, 54*(1), 52–77. https://doi.org/10.1080/0020174X.2011.542943

Gruen, L. (2021). *Ethics and animals: An introduction*. Cambridge University Press.

International Fund for Animal Welfare. (2022). What you should know about sea turtles | IFAW. https://www.ifaw.org/au/animals/sea-turtles?ms=AONDC230001102&cid=7013k000001Ck-Kq&gclid=CjwKCAjw4JWZBhApEiwAtJUN0AdtCxWSzLJ72uZOIRYU6U-9uZ07tnT3v-VhgihEk2Aak3pRhjRgbuRoCdOIQAvD_BwE

Kymlicka, W. & Donaldson, S. (2014). Animals and the frontiers of citizenship. *Oxford Journal of Legal Studies, 34* (2), 201–219. https://doi.org/10.1093/ojls/gqu001

Lockwood, J.A. (1987). The moral standing of insects and the ethics of extinction. *The Florida Entomologist, 70*(1), 70–89. https://www.jstor.org/stable/3495093

Mann, C. (2018). *Vystopia: The anguish of being vegan in a non-vegan world*. Lulu Press.

Pavid, K. (2019, May 6). The world is in trouble: One million animals and plants face extinction. *Natural History Museum*. https://www.nhm.ac.uk/discover/news/2019/may/one-million-animals-and-plants-face-extinction.html

Rowlands, M. (2012). *Can animals be moral?* Oxford University Press.

Ryder, R.D. (2011). *Speciesism, painism and happiness: A morality for the twenty-first century*. Andrews UK Ltd.

Safina, C. (2015). *Beyond words: What animals think and feel*. Henry Holt and Company, LLC.

Schönfeld, M. (1992). Who or what has moral standing? *American Philosophical Quarterly, 29*(4), 353–362.

Singer, P. (1977). *Animal Liberation: A new ethics for our treatment of animals*. Avon.

Sonnenschein, L. (2022, July 30). Tiny turtle pooed 'pure plastic' for six days after rescue from Sydney beach. *The Guardian*. https://www.theguardian.com/environment/2022/jul/30/tiny-turtle-pooed-pure-plastic-for-six-days-after-rescue-from-sydney-beach

11

MORAL STANDING AND ANIMAL WELFARE

This chapter returns to the complex issue of an animal's moral standing being shaped by the socially constructed classification of that animal. With the extension of factory farming to increasingly include non-vertebrate animals such as edible insects and cephalopods, both vertebrate and non-vertebrate animals are now potential commodities for human consumption. Some specific moral and ethical concerns are highlighted in relation to the farming, selling (legal or illegal) and consumption of vertebrate and invertebrate animals, but this is by no means an exhaustive list of the wrongs that many animals experience at the hands of humanity as the following examples highlight.

"1.3 million farm animals dead due to climate change" claimed news headlines as chickens, pigs and cows, all confined in industrial-sized farming systems with no hope of evacuation, died in the British Columbian (Canada) floods and a preceding heat dome event during 2021. (Hill, 2021)

"Heatwave will cause more mass deaths at chicken factory farms, animal charities warn" stated another news item as United Kingdom's unprecedented August 2022 heatwave saw millions of chickens "cooked alive" in intensive farming (Dalton, 2022).

As extreme global weather patterns wreak havoc, the moral and welfare implications of warehousing large numbers of animals inside factory farms where there is no realistic way to evacuate them are gaining considerable attention—although not always for welfare reasons alone. Perceiving factory-farmed animals as property means the deaths of tens of thousands of chickens, pigs and cows in floods or heatwaves are positioned as insurance issues and economic loss rather than the moral welfare issue of death on an unimaginable scale.

When the pain inflicted on farmed animals is given a face, a name and a life story, it becomes harder to ignore. A single animal has the potential to become an ambassador to enact change, one animal at a time, thus bypassing the psychic numbing that can immobilise humans or the cognitive dissonance that causes them to turn away. Tootsie the chicken from Edgar's Mission is one such ambassador of change, with the capacity to put a face to the millions of factory-farmed chickens who are born to die deliberately or indirectly at the hands of humans.

This is Tootsie's tale, reproduced with permission from Edgar's Mission as she stares quizzically into the camera in Figure 11.1 and enjoys life to the fullest at the sanctuary.

DOI: 10.1201/9781003298489-12

Tootsie's Tale, as Told by Edgar's Mission

There is no doubt dear Tootsie smelt the day we first met. Although it wasn't so much a smell that she could claim as her own, but rather a smell that had claimed her. Claimed not only her body, but so much of her life. All of it, to be exact, until the day liberty became her. It was an acrid smell so pungent and stale that it seared our eyes and sored our hearts. A smell of which we are all too familiar. A smell that is the hallmark, and sadly the heart mark, of every sweet and gentle hen who has spent her life in a battery cage★ pumping out egg after egg. Something Mother Nature never tasked her with, but, alas, something humankind has. And has wretchedly done so to the detriment of these innocent birds through selected genetics. It is a smell well known and felt by all those who have cradled a rescued battery hen★ in their loving arms. And although it is a smell we shall never forget, we trust with love, kindness and a new life (and smells) ahead of her, dear Tootsie surely will. For that life and its putrid smell is now where it belongs—behind her.

The term "spent hen" is an industry term used to describe a laying hen who has reached the "term" of her "productive" life, at which point she will be trucked to a place where things never end well for any animal. It is more than an ironic twist that it too describes her circumstance of having "spent" her short, impoverished life in forced servitude in a barren wire cage. By what circumstance Tootsie escaped her former captive life and landed in the front yard of a kindly Good Samaritan (chickens are known to choose their friends wisely), we do not know. But what we do know is this.

Tootsie, complete with her inquisitive mind and itching-to-go feet, is set to make the most of both. It has been our solemn promise to her that this, her nirvana, awaits. She shall be free to feel the sun's warm caressing rays on her back, to scratch about in the soil with all of the industrious sense of wonder a chicken on a mission can muster, to avail herself of a private nest in which to lay her eggs, and to dine on a feast of delicious and varied snacks and treats. But the greatest treat of all we have in store for dear Tootsie is to smell just what kindness is.

Inhale deeply our lovely, inhale deeply…

★The term battery cage and battery hen refer to the cage system in which the majority of hens confined for laying eggs in Australia are kept. The cages are stacked one upon the other resembling the cells in a large battery. There is no nesting material or even a nest in these cages with their wire-sloping floors. Floors so designed to facilitate the eggs rolling forward for ease of collection, but most certainly not for ease of the hens.

Reproduced with permission from Edgar's Mission Sanctuary, Australia.

Chickens demonstrate many of the concepts fundamental to morality, including cooperation, empathy and justice. When given the opportunity to range freely, they establish structured social relations and friendships, often foraging alongside others and sharing dust baths. They display affection, tenderness (caring for weak or sick chickens), altruism (roosters will tell chickens if they have found a particularly enjoyable food) and grief. They form bonds with each other and other creatures, including dogs, cats,

Figure 11.1 Tootsie tilts her head to one side as she stares at the camera. Image supplied by Edgar's Mission and reproduced with their permission.

horses, goats, ducks and humans, ticking all the boxes for moral standing. However, when it comes to their treatment, it is the humans who fail to demonstrate morality.

Further evidence of insurance and economic demands overruling animal welfare issues is evident in the live export of animals mentioned in an earlier chapter. "Some 200,000 animals trapped in Suez Canal likely to die", stated a news headline as an animal welfare tragedy of epic proportions slowly unfolded across one week in March 2021 (Gherasim, 2021). Sixteen vessels remained motionless behind the *Ever Given* cargo vessel wedged across the Suez Canal, placing the lives and wellbeing of 200,000 live animals trapped inside blistering cargo containers with dwindling food and water in jeopardy. Transporting live animals has huge ethical and welfare implications, even when things go to plan, but the practice persists because of the economic benefits and the cultural necessity to eat meat from animals killed in that country. Death by drowning as ships capsize or sink or death by dehydration and starvation when ships are stranded at sea increases the suffering of these doomed animals.

The mandate to prioritise economic benefits over welfare considerations was evident in a political leader's response to another live animal export disaster in the early 2000s. Over 57,000 Australian sheep were condemned to remain at sea in a "floating hell" for three months after their country of destination refused to accept them due to health concerns and other nations turned them away. The Australian Prime Minister at the time responded in a way that unequivocally positioned the animals as commodities rather than living, feeling entities:

I deplore cruelty, any ordinary human being would and does. But we have to keep these things in perspective, we have to remember that you are talking about a very valuable economic asset.

(Radio 5DN Interview, 2003, cited by Wright & Muzzati, 2007, p. 141)

THE FIVE FREEDOMS OF ANIMAL WELFARE

Welfare assessment in captive animals has benefited from growing interest, research and acceptance of animal consciousness, sentience, emotion and suffering, but there is still a long way to go. The Five Freedoms, an evidence-based framework to contextualise the key aspects of animal welfare, evolved from a 1965 British parliamentary enquiry into the welfare of animals in concentrated production environments (RSPCA, 2023). Starting with the basic right of farmed animals to be able to stand up, lie down, turn around, stretch and groom, a decade later it was extended to include both the physical and mental needs of farmed animals, including the avoidance of fear and distress and the ability to express normal behaviour. Five freedoms emerged to assess an animal's welfare:

1. Freedom from hunger and thirst.
2. Freedom from discomfort.
3. Freedom from pain, injury or disease.
4. Freedom to express normal behaviour.
5. Freedom from fear and distress.

In 1994, this was reformulated into five domains to allow a distinction to be made between the physical and functional factors affecting animal welfare and the effect these factors had on the animal's overall mental state. Animal welfare pioneer David Mellor's Five Domains Model for animal welfare assessment, originally developed in 1994 and updated in 2001, 2004, 2009, 2012, 2015, 2017 and 2020, demonstrates the continuum of thought that has evolved to incorporate growing knowledge around the subjective experiences of animals. The original model from 1994 placed an emphasis on assessing and grading the negative impacts on sentient animals used by humans in research, teaching and testing (Mellor et al., 2020). This moved the emphasis from the welfare implications for the animal related largely to the experimental manipulation itself to a broader consideration of the wider conditions and their interactions that could have an additional and cumulative negative burden on the animal. The five domains were (1) nutrition, (2) environment, (3) health, (4) behaviour and (5) mental state. The first three domains focused on the internal state of the animal and any disturbances in these areas. The fourth domain drew attention to external constraints and disturbances such as confinement, presence or absence of other animals (for example, mice can suffer in isolation) and the impact of humans (for example, handling during an experiment). In 1994, the focus was on animal welfare compromise and providing a basis to qualitatively grade the severity of any negative experiences. Animal research ethics applications continue to place a mild/moderate/severe assessment of the suffering burden of every aspect of a study, with the cumulative burden then juxtaposed against the justification in terms of human benefit for imposing this burden on the animals. In most cases, the perceived benefit to humans outweighs even the most severe burden on the animal. The first four domains were largely objective and strongly impacted by external factors with a flow-on effect on the fifth domain, which initially was evaluated in terms of the suffering experienced by the animal in relation to thirst, hunger, anxiety, fear, pain and overall distress.

The concept of distress provided a conduit for consideration of other pertinent subjective experiences as the legitimacy of focusing on animal affective states gradually moved from ridicule to broader acceptance among animal welfare scientists. Survival-critical concerns such as pain, thirst, hunger, weakness, sickness and debility were joined by experiences generated through brain processing of sensory inputs. The animal's perception of their external situation introduced a range of subjective experiences for welfare consideration including frustration, anger, helplessness, loneliness, boredom, depression, anxiety, fear and panic. By the start of the twenty-first century, another fundamental welfare issue was raised. Minimising negative experiences and suffering was only the beginning. Animal welfare could not truly be considered beneficial to an animal without including positive experiences that facilitated the subjective experience of comfort, pleasure, interest, attachment, confidence and a sense of being in control.

The 2015 revised Five Domains of Animal Welfare allowed systematic and structured assessments of animal welfare—including welfare compromise and enhancement—among research animals, farmed animals, animals used in sport, animals kept in zoos, animals used in therapy, terrestrial and cetacean (whales, dolphins and porpoises) working animals and in legal court cases to assess suffering and animal cruelty (Mellor, 2016). By 2020, Domain 4 had been renamed from Behaviour to Behavioural Interactions in recognition of the effects of reduced or enhanced expression of an animal's agency to interact with their environment, other animals and humans (Mellor et al., 2020). The first three domains remained focused on the need for an animal to achieve internal stability of physiological mechanisms fundamental to survival, but stability in Domains 4 and 5 required subjective responses to situation-related factors.

TOOTSIE AND THE FIVE DOMAINS

Tootsie's life story demonstrates the benefits of transitioning from negative to positive experiences in Domain 4. Chickens have the capacity to form close relationships with other chickens and their interactions with the environment. They learn where they fit in their protective social group and can memorise and recognise the faces of over one hundred other chickens (DeMello, 2012). They can communicate with each other, be it conveying to other members of their flock via alarm calls that a predator is approaching by land or by air, or a rooster gaining popularity with the females by taking them to a delicious food find. They scratch and forage, have dust baths together, and generally enjoy life.

In her early life, Tootsie was fed and watered, guarded from predators in a wire cage with many other chickens and kept healthy within legal requirements, often through antibiotics introduced into her feed. This life met the stipulated requirements of Domains 1–3. However, Tootsie lacked meaningful interaction with her environment and other chickens. She was surrounded by other chickens but was not part of a socially defined flock. Tootsie's environmental interactions were also inadequate. She was fed dry pellets, instead of scratching and foraging for tasty treats with different textures and tastes. She had no need to warn her group members of predators because she was locked in a metal cage that distorted her feet and ensured her faeces dropped away for ease of cleaning. Her chicken curiosity had no outlet and her chicken need for nature was ignored. Tootsie lacked agency in a life that offered no outlet for the positive experiences and emotions to remove fear and distress that were required to satisfy Domain 5. Whether Tootsie also felt negative feelings of frustration, anger, helplessness, loneliness, boredom, depression, anxiety, fear and panic is open to conjecture, but any such feelings would have gone unnoticed among the mass of anonymous chickens sharing her experiences. What can be surmised is that Tootsie had never experienced the positive subjective experiences of comfort, pleasure, interest, attachment, confidence and a sense of being in control until the fateful day she entered the safety of Edgar's Mission.

Applying the Zoological Emotional Scale in Figure 11.2 to Tootsie's life demonstrates where change was enacted in the friend domain for the improved welfare of chickens in general. Tootsie became an individual through emotive images and narratives on social media, thus achieving friend status and the attribution of positive emotions. Individualising Tootsie's life story promoted change for one chicken at a time by circumventing feelings of psychic numbing that promote inertia.

Factory farming of domestic animals and poultry has become essential to meet increased human consumption needs. The juxtaposition of Tootsie's early life in a cage with her life at Edgar's Mission underscores some of the welfare implications associated with a method that removes individuality and meaning from the animals who are born to die for human demand. Graphically depicting the emotional distress and degradation of Tootsie's captive anonymous friends can be insufficient to overcome the unpleasant feelings of cognitive dissonance when forced to choose between the love of chicken flesh, chicken nuggets and eggs and concern over faceless chickens. However, meeting Tootsie, or simply hearing her story, opens the door to change, one animal and one person at a time.

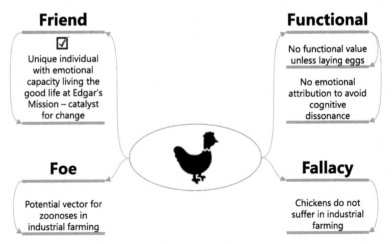

Figure 11.2 The Zoological Emotional Scale applied to Tootsie compares her current life as an individual at Edgar's Mission and her past life as a battery hen.

FACTORY FARMING – ONGOING CONCERNS

Reports on the iniquities of factory farming, live animal export and extreme weather events present death on an unimaginable scale and can trigger the protective mechanism of psychic numbing. For much of the time urban dwelling people are sheltered from animal suffering and death at this level. The "ag gag" laws of some countries limit or prevent the recording of operations at commercial agricultural facilities and the release of any recordings to the public, meaning any footage obtained is illegal and reliant on social media and personal contacts to disseminate. Graphic stories still circulate, including chickens deprived of food and water to bring on a forced moult and increase their already unnaturally enhanced egg-laying capacity; sows contained in breeding crates for the whole of their breeding life; and calves torn from their mothers and death of superfluous offspring to facilitate an endless supply of cow's milk for human consumption.

Research increasingly confirms the common-sense observation that when forcibly parted from their newborn calves, dairy cows demonstrate negative emotional responses such as an increase in visible eye whites (Proctor & Carder, 2015). Nevertheless, cows continue to be separated from their calves immediately or shortly after birth in some countries, based on rationales such as reducing the stress of later separation after bonding occurs; protecting the health of calf and mother if teat damage occurs; and increasing saleable milk from the cow (Neave et al., 2022). By nature, cows choose to live in herds, as Figure 11.3 shows, and the mother seeks isolation to give birth before returning to the group. Strong bonds develop between mother and calf within hours of birth. Mothers nurse their young five to nine times daily for the first few weeks of life and care for them until 6–14 months. The mother–calf bond includes social learning

Figure 11.3 A herd of cows standing together, with four facing the camera. Photograph supplied by the author.

and affiliative behaviours which are never fulfilled when the calf is removed at birth. Some European countries, partly in response to public out-cry and separation images going viral on social media, now provide extended cow–calf contact (Vaarst et al., 2020). In Chapter 4, Clarabelle and Valentine were introduced, highlighting the actions that a mother cow will take to preserve that precious mother–child bond.

The invisibility of deaths and suffering allows some humans to maintain reassuring narratives and economic rationales around the necessity of a mass production style, thus ignoring the consequences on animals and the environment. Researchers Frédéric Leroy and Istvan Praetz (2017) identified inconsistencies in the visibility of animal death depending on cultural variables and ranging from hunter-gatherer, domestic and post-domestic societies. Within this hierarchy, the animal became more objectified as killing was outsourced to slaughterhouses and the living animal de-identified. This de-identification and journey to the slaughterhouse by land or sea increased the potential for negative emotions in animals leading up to, and at point of death, prompting Jonathan Safran Foer to comment in his investigative book, *Eating Animals* (2010) that: "When we eat factory-farmed meat we live, literally, on tortured flesh. Increasingly, that tortured flesh is becoming our own" (p. 143).

Tootsie, Clarabelle and Valentine offer insight into an alternative outcome from the fate of those countless animals bred to live and die in captivity. Describing the before and after of one animal at a time may be a slow way to enact change, but it brings to life the inherent inequities of a system that commodifies animals and privileges human desires. One animal at a time will always be preferable to ignorance.

CASE STUDY–THE PIG

In 1995, a movie about a talking piglet named Babe who bonded with a motherly Border Collie dog and discovered that he too could herd sheep became a box office success. Babe joined Wilbur the pig who was befriended and saved from the dinner table by Charlotte the spider in the classic book and movie *Charlotte's Web* (1973) in successfully but temporarily causing humans—especially children—to rethink the roast pork, ham and bacon on their plates.

Pigs occupy a contested space in human society, with social behaviours reminiscent of companion animal dogs and a reputation for intelligence juxtaposed against descriptors such as gluttony, filth and fat. With sufficient space away from the unnatural confinement of crates and pens, pigs form complex social networks and are fastidious about not soiling the areas in which they eat and sleep. They love to roll in water or mud to cool off because they cannot sweat, contrary to the derogatory saying "sweating like a pig". They also occupy the unsolicited place among mammals of being the only livestock animal bred solely for slaughter. They offer nothing in the way of wool, milk, eggs or labour, putting them in the same league as fish and insects who are commercially farmed. In some countries, pigs are confined to crates, which allow minimal movement, thus forcing them to fulfil the stereotypes of filth and fat when unable to escape their own faeces and vomit. Bred to grow ever fatter until crippled under their own weight or to continuously bear litters of piglets, the end result is always slaughter. Attributing emotional capacity to these creatures destined for death is to invite cognitive dissonance for the humans who consume them.

Piglets, however, can appeal to a person's innate need to nurture. They are cute, smart, sensitive and self-aware resulting in teacup pigs gaining popularity as companion animals. But even teacup pigs grow bigger, especially if they were actually pot-belly pigs malnourished to stay small. Adult pigs are unwelcome in some suburban areas, resulting in relinquishment to shelters and the risk of euthanasia. Emotional capacity is attributed to the cute young piglet and anthropomorphised Babe and Wilbur, but after that life is not so optimistic for adult pigs unless they somehow find their way to a farmed animal sanctuary. This allows them to become an individual with a name, personality and life story and therefore worthy of emotional attribution and moral standing.

Where adult pigs escape and establish wild populations, they are often labelled damaging pest animals and blamed for degrading soil and water, preying on native species and damaging crops. They have also become harbingers of zoonotic disease, with the world pandemic in 2009, colloquially named swine flu, increasing fear of pigs and contributing to the stereotypes surrounding their lives. Once a bacterial, viral or parasitic pathogen jumps the species barrier to infect humans, it is labelled a zoonosis. When African swine fever (ASF) killed an estimated 25% of the world's pig population in 2018, it was reported in terms of the significant impact for the global protein market rather than the deaths of one in four sentient beings (Long, 2019).

Unfortunately, pigs remain associated with negative stereotypes despite factual evidence to the contrary. Pork is one of the most well-known food taboos in Judaism and Islam, although the origins of the pig taboo remain an area of scholarly debate drawing on disciplines such as biology, anthropology, ancient history, mythology, religion and ethology (Lobban, 1994). The word pig has been used as a derogatory term for police since the early 1800s and the term pig-headed is used to describe a human who is stupid, obstinate and unreasonably set in their ways. Visually representing pigs on the Zoological Emotional Scale in Figure 11.4 raises the question of how this misunderstood animal can occupy so many conflicting positions within a human-dominated world. The pig has many erroneous human assumptions and attributions to overcome to establish itself as a worthy emotional being with moral standing. Understanding how the human-assigned social role and perceived emotional capacity for any given pig, or group of pigs, can determine their life and death experiences provides important information during the pre-contemplation, contemplation and preparation stages of change. When applying this framework, each person starts from their own point of familiarity with the animal when considering "why do I think this way and are there other ways of thinking about pigs?" While this may create feelings of cognitive dissonance, it can also start the process of change to address negative stereotypes and outcomes.

Pigs are not alone in being infamous for incubating diseases that threaten human health, livelihoods and economies. In 2021 and 2022, the Avian influenza virus wreaked havoc across the globe. As industrial farms were forced to kill hundreds of thousands of chickens, ducks and turkeys, these deaths were reduced to an economic concern and its impact on employee salaries and company profits. Avian flu occurs in the wild and spreads from country to country as wild birds migrate seasonally. It can spread quickly on factory farms where thousands of chickens, ducks and turkeys are housed in massive industrialised complexes. A bird flu outbreak often requires a mass kill of the entire flock, a process known as depopulation. According to Animal Justice, a Canadian not-for-profit animal advocacy organisation, thousands of birds can be killed at

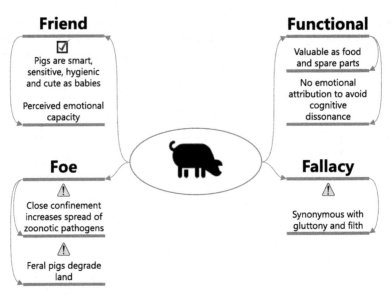

Friend

☑

Pigs are smart, sensitive, hygienic and cute as babies

Perceived emotional capacity

Functional

Valuable as food and spare parts

No emotional attribution to avoid cognitive dissonance

Foe

⚠

Close confinement increases spread of zoonotic pathogens

⚠

Feral pigs degrade land

Fallacy

⚠

Synonymous with gluttony and filth

Figure 11.4 The Zoological Emotional Scale is applied to pigs to identify the contradictory perspectives applied to these domesticated animals.

once through CO_2 gassing (Nickerson, 2022). The detailed instructions describe this method of injecting CO_2 into the sheds as a way of improving personnel safety and reducing biosecurity risks, and almost as an afterthought, maintaining animal welfare.

Epidemics caused by Avian influenza virus (bird flu) associated with industrial poultry farming first appeared in Hong Kong in 1997, while the influenza A virus, also called swine flu, emerged from industrial pig farms in Mexico. The ongoing intensification and industrialisation of food animal production in many parts of the world has seen a concentrated confinement of animals that makes an excellent breeding ground for zoonotic pathogens. Keeping a large number of stressed animals who are often genetically homogeneous to produce higher yields can turn factory farms into "incubators, amplifiers and transmission ports of zoonotic pathogens" (Brozek & Falkenberg, 2021, p. 9251). Mutations and genetic recombination can occur, exacerbating the risk of the development of multi-drug resistance due to overuse of antibiotics on industrial farms, especially for non-therapeutic disease prevention and growth promotion. Once again, it is human wellbeing that is prioritised over the moral implications of confining animals in conditions conducive to zoonotic pathogens.

THE SPREAD AND DIVERSIFICATION OF INDUSTRIALISED ANIMAL FARMING

While considerable attention remains focused on domestic farmed animals such as pigs, cows and poultry, other animals risk confinement for mass human consumption. Human demand for seafood as a source of protein has seen the establishment of fish farms where huge numbers of fish such as salmon are kept in barren environments and at risk of multiple health issues. Insects, described as "mini livestock", are housed by the millions in the growing insect-farming industry. Much of the literature on insect farming relates to the implications for human welfare (for example, avoidance of microbial contamination and parasites) and overcoming human reluctance to eat this food source. The moral standing of these tiny invertebrates raised as a mass production food source in potentially questionable conditions remains undetermined and unregulated (Baiano, 2020). Eating insects has been practised since ancient times and continues to be a cultural norm in some societies. Consumption of locusts, beetles and grasshoppers was approved in the Bible's Old Testament, and in the twenty-first century it is estimated that insects are included in the traditional diets of at least two billion people in over 100 countries, mainly located in Africa, Asia and Latin America. Insects, or derivatives of insects, are also used for farmed animal feed, pets, and circus and zoo animals.

Edible insects can be farmed vertically, requiring considerably less space and resources compared to more traditional mammal and Avian livestock and allowing millions of insects such as crickets to be housed in one facility. Small-scale rearing occurs in many Asian and African regions, where insects are locally sourced and disseminated through local markets. Insect rearing on the mini-livestock industrial scale is more recent and increasing in Western countries as well as Asia and relies on its own core breeding stock. Research among UK insect farmers identified that insects are often dismissed as outside moral obligation because there is limited consensus about what, if any, ethical issues are associated with these animals (Bear, 2021). Unanswered questions about insect sentience including their capacity to feel pain are compounded by limited formal regulations prescribing standards for edible insect production. Their mass killing remains an area of contention and reflects the low moral standing and lack of perceived emotional capacity or even sentience of these living creatures. Insects are generally killed by freeze-drying, sun-drying or boiling, all of which fail to meet criteria related to avoidance of unnecessary distress or harm to a living being, nor the Five Domains of Animal Welfare.

Instructions included in the United Nations Food and Agriculture Organisation publication *Guidance on Sustainable Cricket Farming* (Hanboonsong & Durst, 2020), described as a practical manual for farmers and inspectors, provides instructions on harvesting the adult crickets as if they were produce rather than sentient creatures:

> Once harvested, the crickets can be bagged alive immediately and sold to buyers. However, if the crickets are transported, heat generation in the bags could cause the release of an undesirable amount of histamine, which can cause allergic reactions in some people [...] Other post-harvest options include boiling the crickets for five minutes to sterilise them.
>
> (pp. 15–16)

Given the recent reforms in various countries banning the live boiling of decapod crustaceans such as lobsters, even the possibility that crickets could suffer in a similar way should make any of these harvesting suggestions morally abhorrent. Considerable research on invertebrate capacities has concentrated on adult crickets, resulting in the identification of inter-individual differences in the traits of aggressiveness, general locomotor activity and exploratory behaviour (Lambert et al., 2021). Isolation and deprivation of environmental stimulation and other social experiences during nymphal development accounted for some of these individual traits, suggesting that crickets have an awareness beyond that of a commodity to be bagged alive for transport.

Cockroaches may cause many people to shudder, so the thought of billions of them housed in factory farms can evoke nightmares. When a million escaped from a Chinese farm in 2013, the locals panicked and a massive clean-up operation saw the whole area disinfected. Ground up as part of a healing potion in traditional medicines or fed to livestock as a source of protein, the commodification of cockroaches fails to acknowledge any capacity to suffer nor their importance in natural ecosystems (Millman, 2022). This is compounded by the disgust factor for this resilient creature who can survive blasts of radiation, go for a month without food and remain submerged under water for an hour.

Formally recognising invertebrates as sentient creatures goes some way to protecting their interests, as evidenced in the UK *Animal Welfare (Sentience) Act 2022* that acknowledged the complex central nervous systems in decapod crustaceans and cephalopod molluscs. This recognition did not prevent their human consumption as long as the method of death met welfare standards. Octopuses, described as intelligent, curious and solitary creatures capable of experiencing distress and happiness, gained increased moral standing after the release of the award-winning documentary *My Octopus Teacher*. As with domestic farmed animals like Tootsie, Clarabelle and Valentine, individualising one octopus draws attention to the whole species and potential atrocities committed against them. The very real possibility of these solitary, self-aware creatures being confined in industrial farming complexes to meet global demand for a dwindling product has sparked worldwide controversy.

In sharp contrast to the solitary octopus, honeybees are eusocial creatures who must function as a coordinated whole to survive. They have also shown evidence of self-awareness, personality and the capacity to experience stress. In a mobile equivalent of factory farming, hives of bees are treated as commodities of pollination and shipped over much greater distances than could occur naturally (Ellis et al., 2020). The increase in Californian almond orchards to meet global demand can see an estimated 80% of the total US stock of commercial pollinator bees—approximately two million bee colonies—transported to California every February. Despite inconclusive and mixed reports on the bees' health outcomes, some indications suggest confinement during transportation, variations in temperature, air pressure and vibrations and lack of diversified forage can negatively impact lifespans and promote nutritional deficiencies and the imbalances of oxidative stress (Melicher et al., 2019). Few small beekeepers can now maintain a livelihood from honey and beeswax, especially in industrialised Western nations, making selling pollination services for monocultured crops the primary source of profit from bees. Any perception of the bees' individual and colony needs is thus subjugated to human priorities, irrespective of the bees' self-awareness and capacity to experience negative outcomes.

African American biologist Charles Henry Turner (1867–1923) was fascinated by the behaviour, social organisation, division of labour and collective intelligence of social insects such as bees long before established science viewed insects as having senses and capabilities (Giurfa et al., 2021). Turner did not consider bees as simple machines driven by reflex and spontaneous reactions to environmental stimuli. Behind every bee's decision, Turner asserted, was learning, memory and importantly, individual variability.

Bee expert Lars Chittka (2022) also described bees as smart creatures who must remember colours, scents, shapes and routes when foraging in the "flower supermarket". Chittka believed bees have a "library" of memories that could be selectively accessed. They demonstrated problem-solving techniques and importantly had emotional states evident

Figure 11.5 A bee rests with outstretched wings on a large flower. Photograph supplied by the author.

in their unique behaviours. This evidence, Chittka argued, raised ethical concerns for the treatment of bees in research, conservation and as transportable pollination commodities. For Chittka, as for Turner decades earlier, bees had individual minds, with an appreciation of the outcomes of their own actions and the capacity for basic emotions and intelligence. He described the bee's need to overcome frustration and starvation risk when finding exactly the right flower only to discover many were empty after a competitor had recently drained them. The bee must constantly make decisions which, Chittka contended, were not driven by innate priorities. Individual bees had the capacity to choose differently according to their own predispositions. The bee in Figure 11.5 had many different flowers to choose from in the same locale, ultimately coming to rest on this large flower.

The possibility that bees could have emotional states was explored by British behavioural scientist Mellissa Bateson, US bee scientist Geraldine Wright and their team (2011) using protocols previously adopted to explore emotional states in vertebrates. Both humans and vertebrate animals are more likely to make pessimistic judgements when in an anxious or depressed state. After being trained to associate a sweet reward with a 9:1 mixture of two odours and a bitter reward with a 1:9 mixture of two odours, bees were presented with an ambiguous odour. Half the bees were vibrated to simulate a predator attack and the control group was left undisturbed. The shaken bees appeared to display a pessimistic cognitive bias in their reluctance to do anything about the ambiguous stimulus. Chittka's team found similar results with bees showing positive cognitive bias to ambiguous stimuli immediately following receipt of a surprise reward. Concluding that emotions including fear are survival-related—at times even survival critical—Chittka commented that a large brain was not necessary to achieve this state and basic emotions may be part of the bee's survival tool kit.

In an attempt to raise human awareness of bees as individuals, Chittka's (2022) team marked over 2,000 bees of three species with number tags in multiple colours.

The nests were located on East London's Queen Mary University campus and the bees could forage freely in the flower supermarkets of local gardens, parks and balconies up to 8 km from the hives. As people observed the same bees returning repeatedly to their locations, they came to appreciate them as individuals with unique life stories, individual flower preferences and memories of where those flowers were located. These bees were no longer anonymous commodities of pollination. As noted with Tootsie, Lemonade, Clarabelle and Valentine, individualising animals is more likely to bestow moral standing than horrific statistics of suffering and death.

IMPLICATIONS

The scientific debate as to whether some animals experience emotions is over, and yet discrepancies in how this translates into animal welfare remain. Whether on factory farms or live animal export ships, many animals receive no moral standing in the decisions being made on their behalf.

As well as the reassuring narratives and economic rationales used to allay uncomfortable feelings of cognitive dissonance, psychic numbing can play a role when considering the welfare of vertebrate and invertebrate animals on a massive scale. Psychic numbing becomes a flawed but necessary psychological defence strategy when the sheer scale of suffering and death is too overwhelming to contemplate. Thousands of confined animals drowning or cooking in extreme weather events as global weather patterns change, or problems on live animal export ships can trigger this protective mechanism. Psychic numbing also prevents a person from embarking on the cycle of change as personal responsibility is removed by the volume of death and destruction.

Tootsie and the individualised bees in London gardens reduce the magnitude of ethical welfare dilemmas, grounding the observer in the plight of one animal at a time and thus providing a starting point for change. Applying the Zoological Emotional Scale in an attempt to understand the inequities in the perception of animal emotional capacity and treatment provides further relevant information to transition through the cycle of change. Some of the changes must be enacted at a societal or political level— such as live animal exports—but historically the impetus for higher level change can start from below.

REFERENCES

Baiano, A. (2020). Edible insects: An overview on nutritional characteristics, safety, farming, production technologies, regulatory framework, and socio-economic and ethical implications. *Trends in Food Science & Technology, 100,* 35–50. https://doi.org/10.1016/j.tifs.2020.03.040

Bateson, M., Desire, S., Gartside, S.E. & Wright, G.A. (2011). Agitated honeybees exhibit pessimistic cognitive biases. *Curr Biol., 21*(12),1070–10733. https://doi.org/10.1016/j.cub.2011.05.017

Bear, C. (2021). Making insects tick: Responsibility, attentiveness and care in edible insect farming. *EPE: Nature and Space, 4*(3), 1010–1030. https://doi.org/10.1177/2514848620945321

Brozek, W. & Falkenberg, C. (2021). Industrial animal farming and zoonotic risk: COVID-19 as a gateway to sustainable change? A Scoping Study. *Sustainability, 13*(16), 9251. https://doi.org/10.3390/su13169251

Chittka, L. (2022). *The mind of a bee.* Princeton University Press.

Dalton, J. (2022, August 11). Heatwave will cause more mass deaths at chicken factory farms, animal charities warn. *Independent UK Edition.* https://www.independent.co.uk/news/uk/home-news/heatwave-animal-chicken-mass-death-farm-b2143203.html

DeMello, M. (2012). *Animals and society: An introduction to human-animal studies.* Columbia University Press.

Ellis, R.A., Weis, T., Suryanarayanan, S. & Beilin, K. (2020). From a free gift of nature to a precarious commodity: Bees, pollination services, and industrial agriculture. *Journal of Agrarian Change, 20*(3), 437–459. https://doi.org/10.1111/joac.12360

Gherasim, C. (2021, March 30). Some 200,000 animals trapped in Suez canal likely to die. *EUObserver.* https://euobserver.com/world/151394

Giurfa, M., Giurfa de Brito, A., Giurfa de Brito, T. & de Brito Sanchez, M.G. (2021). Charles Henry Turner and the cognitive behavior of bees. *Apidologie, 52,* 684–695. https://doi.org/10.1007/s13592-021-00855-9

Hanboonsong, A. & Durst, P. (2020). *Guidance on sustainable cricket farming - A practical manual.* Bangkok, FAO. https://doi.org/10.4060/cb2446en cb2446en.pdf

Hill, B. (2021, December 7). 1.3 million farm animals dead due to climate change: What can B.C. do to stop the next catastrophe? *Global News.* https://globalnews.ca/news/8427762/b-c-flooding-kills-650000-farm-animals/?fbclid=IwAR31NWWllENlhieJzkaJZqKHb5gB2zhB56qMp0TlZy91-ysWHypugUpAzQA

Lambert, H., Elwin, A. & D'Cruze, N. (2021). Wouldn't hurt a fly? A review of insect cognition and sentience in relation to their use as food and feed. *Applied Animal Behaviour Science, 243* (October). https://doi.org/10.1016/j.applanim.2021.105432

Leroy, F. & Praetz, I. (2017). Animal killing and post domestic meat production. *Journal of Agricultural and Environmental Ethics, 30,* 67–86. https://doi.org/10.1007/s10806-017-9654-y

Lobban, R.A. (1994). Pigs and their prohibition. *International Journal of Middle East Studies, 26*(1), 57–75. https://www.jstor.org/stable/164052

Long, W. (2019, September 18). One quarter of the world's pigs killed by African swine fever as disease spreads to South Korea. *ABC News.* https://www.abc.net.au/news/rural/2019-09-18/one-quarter-of-worlds-pigs-killed-by-swine-fever/11524134

Melicher, D., Wilson, E.S., Bowsher, J.H., Peterson, S.S., Yocum, G.D. & Rinehart, J.P. (2019). Long-distance transportation causes temperature stress in the honey bee, *Apis mellifera* (Hymenoptera: Apidae). *Environmental Entomology, 48*(3), 691–701. https://doi.org/10.1093/ee/nvz027

Mellor, D.J. (2016). Updating animal welfare thinking: Moving beyond the 'Five Freedoms' towards a 'Life Worth Living'. *Animals, 6*(3), 21–41. https://doi.org/10.3390/ani6030021

Mellor, D.J., Beausoleil, N.G., Littlewood, K.E., McLean, A.N., McGreevy, P.D., Jones, B. & Wilkins, C. (2020). The 2020 Five Domains Model: Including human-animal interactions in assessments of animal welfare. *Animals, 10*(10), 1870. https://doi.org/10.3390/ani10101870

Millman, O. (2022). *The insect crisis: The fall of the tiny empires that run the world.* W.W. Norton & Company, Inc.

Neave, H.W., Sumner, C.L., Henwood, R.J.T., Zobel, G., Saunders, K., Thoday, H., Watson, T. & Webster, J.R. (2022). Dairy farmers' perspectives on providing cow-calf contact in the pasture-based systems of New Zealand. *Journal of Dairy Science, 105,* 453–467. https://doi.org/10.3168/jds.2021-21047

Nickerson, S. (2022, April 29). 1 Million+ Birds Killed on Canadian Farms Amid Bird Flu Outbreaks. *Animal Justice.* https://animaljustice.ca/blog/birds-killed-bird-flu?fbclid=IwAR19V9B5oM3-mhyixDTC9X7FiBGsLb0bq-Kb_yzISQD7sTRleH7pS_7EhNU

Proctor, H.S. & Carder, G. (2015). Measuring positive emotions in cows: Do visible eye whites tell us anything? *Physiology & Behavior, 147,* 1–6. https://10.1016/j.physbeh.2015.04.011

RSPCA. (2023). *What are the Five Freedoms of animal welfare?* RSPCA Knowledgebase. https://kb.rspca.org.au/knowledge-base/what-are-the-five-freedoms-of-animal-welfare/

Safran Foer, J. (2010). *Eating animals.* Back Bay Books.

Vaarst, M., Hellec, F., Verwer, C., Johanssen, J.R.E. & Sorheim, K. (2020). Cow calf contact in dairy herds viewed from the perspectives of calves, cows, humans and the farming system. *Farmers' perceptions and experiences related to dam-rearing systems. Journal of Sustainable and Organic Agricultural Systems, 70*, 49–57. https://doi.org/10.3220/LBF1596195636000

Wright, W. & Muzzatti, S.L. (2007). Not in my port: The "death ship" of sheep and crimes of agri-food globalization. *Agriculture and Human Values, 24*, 133–145. https://doi.org/10.1007/s10460-006-9056-7

12

MORAL STANDING AND HUMAN NEEDS

A perplexing question arises as to whether an animal can achieve higher moral standing when beneficial changes to their welfare are primarily for human wellbeing. Humans have the power to make decisions that affect animals on a small and large scale. High moral standing engenders concern and respect and promotes inclusion of the animal's perspective when making those decisions, while minimal or low moral standing disregards the animal's perspective and can perpetuate their suffering. Animals may be left in morally questionable circumstances when there is limited or no motivation for humans to move to the action phase in the cycle of change. Where motivation for change does arise, but primarily for human benefit, the animal may simply be a fortuitous bystander rather than positioned as a being with moral standing.

In 2020, zoonoses came under the spotlight as the COVID-19 virus infiltrated all corners of the human world. COVID-19, formally labelled SARS-CoV-2, is a coronavirus described as a spill-over from wildlife sources. Once a bacterial, viral or parasitic pathogen jumps the species barrier to infect humans, it is labelled a zoonosis. Humanity's ever closer relationship with animals in agriculture, homes and encroachment on wildlife habitat dramatically increased the opportunity for spill-over of pathogens between animals and humans, leaving many epidemiologists unsurprised when COVID-19 appeared. Up to 15% of all known human pathogens have emerged from zoonotic viruses, and zoonoses represent over 65% of all pathogens discovered since 1980 (Wildlife Conservation Society, 2020). Even before COVID-19 made global headlines, zoonoses were a public health issue of increasing concern.

Animals living and dying in the close confines of factory farms are one means by which pathogens with the potential to jump the species barrier can spread exponentially. Wildlife markets, otherwise known as wet markets, are another recognised source of zoonotic infections. They have come under increasing pressure to cease operations since the COVID-19 pandemic made this animal welfare issue a matter of global human concern. Live animal markets are commonly found in the tropical and sub-tropical regions of the world (Galindo-González, 2022). These markets stock a variety of live mammals, poultry, reptiles and fish, often close to areas selling dead fish, red meat or prepared food. The forced close interactions of species that would never meet under normal circumstances can have unpredictable consequences. In their natural ecosystem and biogeographic region, different species of wild animals interact freely when using their environment's resources. However, concentrating wild species foreign

DOI: 10.1201/9781003298489-13

to each other, including frozen, dead or alive amphibians, snakes, turtles, insects, bats, civets, pangolins, marmots and field rats, alongside the domesticated poultry, dogs, cats, rabbits and pigs, the peri-domestic rodents and birds, and humans, can create a pool of potential zoonoses just waiting to jump species. Some species in particular, including primates, bats, pangolins, civets and rodents, are believed to have high-risk reservoirs of virulent pathogens.

Add the stressful and harsh conditions that compromise the immune systems of these confined animals, and the mixing pot is primed for pathogens to mutate, recombine, evolve and colonise new hosts in these unnatural environments. Images of injured, weak and stressed animals packed into open cages, eating inappropriate diets and defecating where they stood or lay remain a hallmark of the early days of the COVID-19 pandemic. Wild animals kept alive so they could be killed on the spot and delivered fresh to market buyers were often butchered on the same surfaces, exposing carcasses to a mix of blood and other organisms that were subsequently consumed by humans (Wyatt et al., 2022). Described as time bombs poised to explode into human populations, wildlife markets also provided an outlet for illegal wildlife trade, especially parts of animals greatly prized for human consumption, traditional medicines, ornamental use or private collection (Córdoba-Aguilar et al., 2021). Under these circumstances, animals had no moral standing as they were primarily commodities for human usage.

Following the 2002–2004 SARS Co-V coronavirus outbreak marking the first global pandemic of the twenty-first century, consistent warnings from conservationists, epidemiologists and virologists for urgent controls and even the complete banning of live animal markets escalated. Counter arguments of restricted income and access to food for some local communities, with the added threat that trade would then be driven underground (Wikramanayake et al., 2021), consolidated the primacy of human interests and ignored the ethical implications of treating sentient creatures in this way. Certain animals, such as the pangolin, a scaly-skinned mammal and the civet, a small cat-like mammal, have been demonised over the years in the frantic need to apportion blame for medical catastrophes that were not of their making. No animal would choose to be crammed into stacked cages in unsanitary, stressful conditions with probable death as the endpoint. Instead, zoonoses and pandemics highlight the interconnection between animal welfare, human welfare and the natural environment, with the real triggers for pandemics being the amplification loops provided by human society.

CASE STUDY – DEMONISATION OF THE CIVET

Civets are not cats, despite the popular tendency to label them as such. These small mammals may have a cat-like appearance, with a thick, furry tail and expressive eyes, but the similarities end with their pointy snout and their eating habits. There are approximately 15 different civet species, residing predominantly in Asia and Africa. In their natural environment, civets play an important ecological role, from defecating and dispersing seeds to biological pest control. These nocturnal solitary creatures suffer greatly when caged in crowded civet farms or wildlife markets. Civets are trapped and confined because their flesh, their faeces and their perineal gland secretions are of value to humans. No other creature defecates a very expensive novelty brand of coffee when force-fed coffee berries, while scrapings from their perineal glands (near the anus) have also been a prized fixative for perfumes, although synthetics have replaced demand for this bodily secretion. Without freedom, exercise, space and proper food, civets experience infections and stress-induced stereotypic movements as they spin, pace and bob their heads in frustration. Unfortunately, even their unique coffee-producing abilities did not stop thousands of masked palm civets being slaughtered in China as suspects in the 2002–2004 SARS Co-V coronavirus outbreak (Watts, 2004).

As some species of civets dwindle in their natural habitats, they increasingly occupy a complex and contradictory position in human society dictated by human control and needs. When explored within the Zoological Emotional Scale in Figure 12.1, their emotional capacity for positive and negative experiences is either negated by their functional or foe status or anthropomorphised through the selective breeding of paedomorphic features that turn them into an infantilised friend with wild animal origins. Civets have gained popularity as a status symbol companion animal for urban youth throughout Indonesia, resulting in a network of home breeders to cater to the demand (Hooper 2022). Locked in small metal or wooden cages stacked on top of each other, these civets exist in distressing conditions reminiscent of the live animal markets of Asia. All these human-specified roles fail to respect the natural lifestyle of a civet where they play an important ecological role in seed dispersal and pest control. As friend, foe or functional commodity, the civet's positioning can have negative consequences for their welfare.

EXOTIC PET TRADE – ANIMALS AS PROPERTY

The popularity of the civet as a status symbol companion animal highlights another welfare concern that removes moral standing from wild animals. According to World Animal Protection statistics, there are over 17 million exotic pets in the US alone. Many were abducted from their natural wild environment and subjected to long and arduous journeys, often crammed into containers with limited or no food and water (World Animal Protection, 2020). Animals exploited in this way can suffer immensely, with no guarantee of relief at the other end should they survive to be sold as companion animals, for aquariums or entertainment and a lifestyle far removed from their natural requirements.

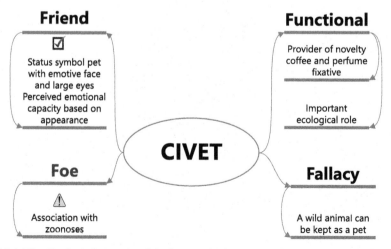

Figure 12.1 The Zoological Emotional Scale is applied to the civet to identify where emotional capacity is attributed to this small mammal.

Whether being sold in a live animal market or illegally trafficked in a tube, animals have no moral standing and all five domains of animal welfare are breached. They can experience disease, injury and functional impairment; deprivation of food and water; environmental challenges including temperature extremes and harmful structures and floors; behavioural or interactive restrictions; and the resultant anxiety, fear, pain and distress that extends into their new life so different to the space and freedom they should have had in the wild. In both live animal markets and exotic pet trade, animals are positioned as property, commodities and resources with no capacity for an emotional life or moral standing. Rather than starting from the premise of animals with moral standing in their own right, the focus is inverted to position humans as having higher moral standing and with it the right to determine treatment—humane or otherwise—of dependent animals. How they determine that treatment can be influenced by where the animal falls in their worldview. When examining the various positions that civets occupy on the Zoological Emotional Scale, despite all civets having the same neural circuits and parts of the brain associated with affective experience and processing, their fate is subject to human culture, historical time period and external events. This outcome is repeated across many different species subject to human domination.

THE MEDICINAL INGREDIENTS OF ANIMALS

In 2007, Marc Bekoff dedicated his book *The Emotional Lives of Animals* to two special animals. There was Pablo, a research chimpanzee who had been darted 220 times, relentlessly subjected to liver, bone marrow and lymph node biopsies, and injected with lethal doses of pathogens before succumbing to an infection associated with these activities. And there was Jasper, a moon bear, confined to a crush cage from infancy for collection of bile. Rescued after 15 years of bile collection, Bekoff described Jasper, shown in Figure 12.2, as "the spokes-bear for hope and freedom. Despite his torture, Jasper forgave" (p. 27).

Figure 12.2 Jasper the moon bear facing the camera and enjoying life as a free bear. Photograph provided by ©Animals Asia Foundation and reproduced with their permission.

Moon bears are gentle individuals capable of expressing joy, curiosity and unique personalities. However, their bodies contain a human-prized substance that has denied many moon bears the opportunity and freedom to express these feelings. The gall bladder of the moon bear stores bile, a product that has a long history of use in some Asian traditional medicines, especially for the treatment of human liver and gall bladder conditions, putting moon bears at risk of exploitation.

According to Animals Asia (2020), an organisation devoted to improving animal welfare across Asia since 1998, bear bile farming imposes relentless suffering on an animal who has the capacity to feel pain and negative emotions. Initially, wild bears were hunted, killed and their entire gall bladder removed. Killing the source of bile added to the scarcity and cost of this product, making it almost inevitable that a more economical alternative would eventually emerge. By the 1980s, bear bile farms met this need by keeping bears alive with constant access to the bile being produced in their liver and stored in the gall bladder (Animals Asia, 2021). Accessing an internal organ like the gall bladder necessarily involves pain. Keeping that access site open to drain bile in unhygienic conditions for the duration of the bear's productive life intensifies the suffering, especially where an infant bear's growing body must contort to fit the metal cage. Denied all five domains of animal welfare, including freedom to express normal behaviours and freedom from fear and distress, the bears' emotional capacity and moral standing is eroded through ongoing physical and psychological suffering.

International travel opportunities and social media have brought the plight of bile bears to global attention, especially the moon bears who have been most commonly exploited and at risk of extinction because of this. First-hand experience of bears experiencing pain and misery inside metal cages can convert an abstract awareness of animal cruelty into a catalyst for wider change. This was the case for both Marc Bekoff and Jasper, and CEO and founder of Animals Asia, Jill Robinson, who tells the poignant story of meeting a moon bear she named Hong while on a trip to Asia in 1993 (Animals Asia, 2018). When viewed through the lens of the Zoological Emotional Scale, it is possible to see how first-hand experience of just one animal in distress can reposition the bile bear from a functional emotionless commodity into a friend whose behaviours on achieving freedom leave no doubt as to their capacity to experience a range of positive emotions.

For people with no first-hand experience of moon bears, their emotive eyes and facial features graphically portrayed in media images and stories can be sufficient to trigger compassion and motivation to support change. Pictures such as Figures 12.3 and 12.4 of the rehomed moon bears relaxing, playing with objects, interacting with each other and enjoying a rich and meaningful emotional life are contrasted with the heart-breaking images of broken moon bears with sad eyes trapped in crush cages for decades. Social media and the world wide web have been instrumental in disseminating the suffering of animals such as moon bears, creating an illusion of friendship with an animal never met.

However, perspectives vary for other people whose economic livelihood—and that of their families—may depend on maintaining bear bile farms. Moon bears become a functional commodity, fundamental in the chain of production and distribution of bear bile, and who must necessarily be de-individualised and perceived as devoid of emotional capacity and moral standing. To perceive them as more than an object delivering bile is to risk the unpleasant mental conflict of cognitive dissonance when having to

Figure 12.3 A group of moon bears relaxing and playing. Photograph provided by ©Animals Asia Foundation and reproduced with their permission.

Figure 12.4 Baruffa the moon bear lies upside down on a swing. Photograph provided by ©Animals Asia Foundation and reproduced with their permission.

choose between maintaining an income and acknowledging the suffering of a sentient creature. In this case, first-hand experience fails to trigger perceptions of emotional capacity, despite the imprisoned moon bears having the same neural substrates that facilitate emotional capacity in the joyful, playful and unique released bears.

Moon bears are not the only victims of human exploitation in the global demand for traditional medicines. Despite many plant-based alternatives, some animals remain subject to cruel practices such as confining lion and tiger cubs in cramped, dirty cages, away from their mothers, until they are killed for the medicinal properties of their young bones. In the oceans, sharks fall prey to demand for both traditional medicine of oil and cartilage, and equally concerning, the showing of wealth and status associated with shark fin soup. Removing fins from a live shark and throwing the crippled animal back into the sea guarantees a fast death by predators or a slow death through incapacitation. And the pangolin, whose need for more extensive protection was raised as long ago as 1938 in an article about their scales having perceived medicinal powers, still retains the dubious honour of being the most trafficked mammal in the illegal wildlife trade (Sexton et al., 2021). Pangolins are the only mammal covered entirely in scales and it is these scales and their flesh that make them irresistible to humans. In the wild, they roll into a tight ball when threatened, exposing only their tough scales for protection, but that is insufficient to stop human poachers many times their size.

ANIMALS IN RESEARCH

Bile, bones, scales and fins are not the only animal parts that humans covet. Animals used in research lose their autonomy and in some cases their life, to promote human wellbeing.

In 2015, the Oxford Centre for Animal Ethics released a paper entitled *Normalising the unthinkable: The ethics of using animals in research* (Linzey et al., 2015). The report succinctly critiqued the routine abuse of sentient animals by subjecting them to harm, pain, suffering and stressful confinement. They questioned the normalisation and moral anthropocentrism that human interests required so many experiments and that human needs, wants or desires should have absolute or near absolute priority in moral calculations.

Quantifying the number of animals used in global research can be difficult, especially where some smaller rodents, fish, reptiles and amphibians and the young they produce as part of a research protocol may be excluded from some reportable statistics. As these groups comprise the bulk of research animals, their omission necessarily misrepresents an overall perspective. The creation of genetically modified animals to help produce better models of human disease further distorts statistics and animal welfare, as genetic manipulation can be unpredictable with unanticipated negative consequences for the animals. In most cases, animals used in research are sentient and able to experience suffering, distress and harm, increasing the risk of cognitive dissonance if handlers are unable to de-animalise their research subject as they perform some of the more invasive procedures.

Most justifications for using animals in research are grouped under the anthropocentric perception that human interests will always override animal interests. However, increasingly the debate that sentience should also invoke moral rights, not least being that animals have some intrinsic worth in themselves as feeling beings, is raising ethical concerns. To accept sentience in animals infers there can be no rational reason for excluding them from at least some of the basic moral considerations afforded a human. As with infants and children, there is a strong case to be made for positioning animals as requiring special moral concern in conjunction with assessment of welfare implications, especially in an anthropocentric world where animals are not given the opportunity to give or withhold consent. They are vulnerable creatures who cannot vocalise their own interests as humans do, leaving it to humans to recognise and act on universal behavioural indicators of distress.

Deciding what to do with animals who remain alive at the end of a research study can be challenging. The Australian *Code of Practice for the Care and Use of Animals for Scientific Purposes* (National Health and Medical Research Council, 2013) provides one suggestion in Section 3.4.2: "Opportunities to rehome animals should be considered wherever possible, especially when the impact of the project or activity on the well-being of the animal has been minimal". The New Zealand Anti-Vivisection Society campaigned for years for mandatory rehoming of animals from research, while Humane Research Australia has lobbied to have dogs and cats used in research routinely rehomed rather than euthanised. However, the endpoint of many studies is death for the subjects, especially genetically modified small rodents who cannot be rehomed. Animal Ethics Committees at major universities and research facilities routinely sign off on studies that end with the humane killing of countless test subjects. They assess Incident Reports should an anonymous numbered animal be found dead, or unwell and follow the Code to ensure the circumstances of an animal's brief life and death are in compliance.

A growing awareness of animals as having moral standing and therefore worthy of ethical consideration has seen the emergence of grassroots movements dedicated to rehoming research animals where possible at the completion of a study. Where research protocols do not specify an endpoint of humane killing, not-for-profit organisations such as Liberty Foundation Australia provide a pathway for retired research subjects into private homes, where there is a focus on their selfhood, freedom of choice and creating long-lasting bonds with humans and other animals. Since commencing operations in 2017, Liberty Foundation Australia has successfully facilitated a new and extended life for more than 650 animals.

Liberty Foundation Australia adopts a strong social media presence to engage the general public in its work and find homes for the animals being retired from research. Each animal is individualised by name and description, often in stark contrast to their status in research where they perform an anonymous role as part of a business model. As with rehomed domestic farmed animals, names provide status for an animal, even though under Australian law they are still deemed property.

Liberty Foundation Australia's 2021–2022 annual update lists the names of animals who have "each left little footprints in the hearts of us all" as they move to their new homes, including guinea pigs Peanut and Nugget, rat brothers Bo and Arrow, Beagle friends Trudi and Princess and rabbits Jill and Bosley. Regular Facebook posts and evocative images continue to individualise those animals still seeking a home and a future, such as two little female mice named Champagne and Gypsy who "…love digging and foraging, running, climbing, making nests and hanging out together in their new naturalistic enclosure". This description is typical of the way Liberty Foundation focuses on the rehomed animals' future outside of the research facility, a strategy that founder and director Paula Wallace describes as necessary to provide a sense of agency for those supporters seeking to become involved. To contemplate the lives of countless, research animals that will never be rehomed can lead people to psychic numbing and a sense of powerlessness. To focus on a few personalised animals at a time provides a sense of purpose and the belief that a difference can be made in the emotional lives of these unique individuals.

Banjo the rat provides a case study of evolving emotional capacity and the attachment relationship he chose as an autonomous sentient being. Banjo and his brother rat entered foster care together in the first step to being rehomed as a pair. When Banjo's brother died and he was relocated to await another suitable buddy, Banjo took matters into his own paws and bonded with one of the adolescents in the household. Banjo's joy at spending time with his human was evident through play, trust, grooming, cuddling, shared quiet times and cross-species emotional contagion as he learned to sense and respond to the adolescent's emotions. Nobody could doubt Banjo's emotional capacity to live life to the fullest right up to the day he died of old age in the arms of his grieving human. Banjo showed no obvious ill effects from his previous life experiences. Instead, he demonstrated a capacity to live in the moment without being burdened by the ruminations to which humans with their perceived superior cognitive abilities often succumb. Living in the moment, a component of Marc Bekoff's concept of animal spirituality, was evident in many of the rehoming examples provided by Liberty Foundation Australia.

There were the two Beagles, Molly and Nala, who displayed an awakening of spirit within hours of commencing their journey from a research facility to their new home. Life for these dogs commenced in a puppy farm, before relocating to a research facility where they remained for several years. At eight and nine years of age, they were ready to embark on the next stage of their lives. While the facility animal technicians familiarised them to human presence and assisted them to overcome much of their timidity, it was a huge transition for these dogs to be picked up by strangers and set out on a lengthy trip in the back of a van. Their subdued body language as they trembled in their travel crates spoke volumes. It was during a toilet break at a grassy field that an amazing moment occurred, as Paula described:

> They were walking around having a sniff for a little while, and then they just decided to lay down and roll in the grass. I knew these girls were timid, and I knew they were frightened, but they just chose to be in the moment. One of them looked at me, and she had this look of pure joy on her face. Her face was relaxed and it was like, Gee, this is nice, isn't it. I'm just going to enjoy it while I can. They weren't thinking, what's going to happen to me, where are you taking me, they were just thinking how wonderful it was to be rolling in the grass in the sunshine. I saw in that moment their spirit was still alive.

Figure 12.5 captures the moment when Molly and Nala experienced the joy of grass during their car trip to a new life.

Humans can never know what an animal is feeling as they roll around in sheer abandon, but Paula's description suggested the experience had meaning for those dogs at that point in time. It promoted their wellbeing in a way that did not require language or interpretive cognitive functions, often cited as evidence of human exceptionalism. Irrespective of their earlier life experiences, these dogs retained an emotional and spiritual spark that was activated at that moment.

Mice provide another example of apparent joy in the moment. Once removed from the standard cages stacked on shelves in light and temperature-controlled laboratories and rehomed in the naturalistic enclosures recommended by Liberty Foundation Australia, distinct and striking behaviour changes can be observed such as digging and climbing to mention a few. Paula described:

"It's as if the mice are singing as they busily go about their lives in their new environment, they seem joyful to be able to engage in behaviours that many people probably don't know are natural habits for mice. Even after decades of breeding these mice

Figure 12.5 Molly and Nala enjoy the touch of grass during a rest break in their car trip to a new life. Photograph by Paula Wallace, Liberty Foundation Australia and reproduced with their permission.

for use in laboratories, it's amazing to see how naturally they begin to dig, forage, run, gnaw and explore, none of which they can usually do in research facility housing".

In fact, according to Paula, they appear to be the most joyful of animals who, when left to themselves, are so much fun to watch as they come out at night and become distinguishable individuals with unique personalities and behaviours.

CASE STUDY – DYLAN, DOLLY, DAISY AND DREW

Female mice Dylan, Dolly, Daisy and Drew were rehomed by Liberty Foundation Australia in 2020. Described as friendly and inquisitive girls who love hanging out together, their lives were irrevocably changed once perceived as unique individuals worthy of a second chance at life (Figure 12.6). When viewed through the lens of the Zoological Emotional Scale, the contemplation and preparation stages of change were instigated within the fallacy and functional domains, respectively. Both contemplation and preparation required a change of mindset to allow conversion of four mice from commodities with low moral standing into lovable friends for those humans prepared to devote time to accepting the unique needs of these industrious little animals (Figure 12.7).

THE ETHICS OF SPARE PARTS

When a tiny pig named Babe shot to worldwide fame in the 1995 movie that carried his name, adults and children alike were entranced by his human-like behaviours and characteristics. Rather than meeting the fate of other pigs herded into a big truck that they believed was taking them to a pig paradise so wonderful that no pig ever came back, Babe achieved almost human-like status as he talked his way off the Christmas dinner menu and into a new career as a sheep-pig.

Figure 12.6 Dylan, Dolly, Daisy and Drew's naturalistic enclosure, with two of the mice popping out from a tube and one mouse under a rope basket. Photograph by Paula Wallace, Liberty Foundation Australia and reproduced with their permission.

FALLACY	CONTEMPLATION STAGE

•There is a perception that because research mice look the same, they are the same. While their suffering must be minimised and environmental enrichments provided to meet welfare standards, they are not perceived as unique, emotional individuals. To perceive them as such risks internal conflict when manipulating and killing them as functional commodities.

FUNCTIONAL	PREPARATION STAGE

•As part of a larger cohort, Dylan, Dolly, Daisy and Drew had designated numbers and were manipulated and monitored according to the research protocol. The function of many mice is to be born, experimented on and humanely killed with no option for rehoming.

FRIEND	ACTION STAGE

•Dylan, Dolly, Daisy and Drew reached friend status when they became individuals. Movement through contemplation and preparation for change in the fallacy and functional domains allowed these mice to be perceived as individuals with unique personalities and emotional capacity. Friend status for these mice could not be achieved without human intervention and the power to elevate their moral standing.

Figure 12.7 The Zoological Emotional Scale shows how perceptions of these mice moved from the fallacy domain to the friend domain.

More than two and a half decades later, a nameless pig also achieved almost human-like status after being genetically modified to become a human heart donor. In 2022, a US man became the first human to receive a heart transplant from a pig–human hybrid. Hailed as a medical breakthrough, the procedure nevertheless re-opened ethical debates around genetically modifying any animal to serve as human spare parts for xenotransplantation.

Julia Baines, science policy adviser for People for the Ethical Treatment of Animals (PETA UK), responded to this medical breakthrough with the observation in

New Scientist that "Using animals as warehouses for spare parts is morally wrong. Pigs are sentient, complex, intelligent individuals that, like any animal, shouldn't be raised and raided for their organs. People who need a transplant need a human organ [...]" (Baines, 2022).

A strong belief in self-importance and exceptionalism coupled with the means to enact change at a global level has seen humans test the boundaries of ethical behaviour towards animals for centuries. According to Melanie Challenger (2021), humans have increasingly become agents of evolution and change with powers far exceeding those of natural selection. Selectively breeding a pig–human resulted in considerable debate around the ethics of xenotransplantation, but largely focused on the consent of individual human recipients and the potential transmission of new infectious diseases to the general population. Omitted from these considerations were the ethical implications of breeding a pig whose sole purpose in life was to provide spare parts for the dominant species. This moves the role of genetically modified animals from providing a spiritual conduit to the natural world to providing life itself.

Xenotransplantation is not a new concept, with the xenotransfusion of lamb blood to humans dating back to the 1600s (Lu et al., 2020). A rabbit kidney was transplanted to a human in 1905, and trials of kidneys, hearts and livers transplanted from nonhuman primates to humans are documented from the 1920s to the 1990s. In the 1980s, a premature infant girl survived 20 days with an adolescent baboon heart, while six patients receiving baboon kidneys lived between 19 and 98 days in the 1960s (Johnson, 2022). Moral reservations about using nonhuman primates coupled with an unacceptably high risk of zoonotic infections saw research endeavours shift towards genetically modifying pigs to provide virus-free, human-compatible organs with reduced risk of catastrophic rejection and infection. Pig organs match human size requirements, with the added bonus of a plentiful supply from large litters, short maturation periods and less public concern about killing pigs compared to primates. Research and farmed animals have long been modified to meet human needs, whether that be research mice with or without specific genes to achieve a study protocol or chickens bred to grow large so quickly that their legs give way. However, genetically altering the smart, social pig to create organs compatible with humans raised ethical qualms that were countered with arguments that pigs globally were bred, transported and slaughtered to provide their flesh for human consumption.

Marc Bekoff (2001) commented that when nonhuman animals are set apart from, and below humans, the interconnectivity and spirit of the world will be lost. Pigs bred to provide organs for humans are undeniably set below humans while simultaneously imbued with human genetic components to make them compatible with humans. Spirituality is an attribute jealously guarded by humanity, and yet one must wonder where the intelligent, sociable, emotional pig, modified to carry human genes, would fit on this human-centric perception of spirituality. The human–pig must necessarily retain its stature as less than human in both emotion and spirituality, for to do otherwise is to struggle with the cognitive dissonance of tampering with nature to service humans.

PARKINSON'S DISEASE RESEARCH REVISITED

Ever since my mother's struggle with Parkinson's disease, I have remained trapped in the murky middle of morality and animal research subjects. It seems animal research can be beneficial to humans, and without the sacrifice of many anonymous animals, the secrets of dopamine and Parkinson's disease may have remained unanswered in my mother's lifetime and left her trapped in the rigidity and tremors of this debilitating neurological condition.

In 2002, results from a British investigation into the use of marmosets in neurological research, including Parkinson's disease, at Cambridge University shocked the research community. Hundreds of monkeys had been confined to barren cages their whole lives, and deliberately brain damaged in a bid to simulate a marmoset model for human neurological diseases (Linzey & Linzey, 2018). Tests specific to Parkinson's disease saw the monkeys shut in tiny Perspex boxes for up to an hour at a time to see how often the brain damage previously inflicted caused them to rotate. They were also injected with amphetamine or apomorphine to make them rotate faster or in the opposite direction. Inflicting brain damage through cutting or sucking out parts of the brain or injecting toxins resulted in pain, distress, bleeding, swelling and bruising, fits, vomiting, tremors, failure to eat or drink, abnormal body movements or loss of movement, balance or visual disturbances, memory impairment and lack of self-care. Once again, the suffering of these marmosets and the countless rabbits and mice who died to gain a greater understanding of Parkinson's disease demonstrates moral anthropocentrism, the assumption that human needs, wants or desires should take precedence in moral calculations over the use of animals.

As the world came to terms with the global COVID-19 pandemic, countless millions of research animals were enlisted in the race to find a vaccine. There is a bitter irony in the sacrifice of these lives in an attempt to control a pandemic potentially arising from humanity's mistreatment of animals and the environment in the first place.

IMPLICATIONS

There is evidence that attempting to protect wildlife by using ethical arguments about species decline and poor welfare has the potential to alienate those humans not ready, or not in a financial position to change their behaviours. However, when the implications of failing to change have a greater negative impact on humans than enacting change, a different outcome is possible. Once personal fears bolster ethical outrage, change becomes more likely, as evidenced in the push to close wildlife markets and control wildlife trafficking in the wake of the COVID-19 pandemic.

The increased presence of zoonotic pathogens in human populations has accelerated international concern and outrage, moving sufficient individuals, organisations and even governments from the contemplation stage to the action stage of behavioural change. The COVID-19 pandemic made the threat personal, adding to the voices of the dedicated campaigners who had long sought to put an end to these animal welfare issues. Warnings that the cost of wildlife trafficking, when viewed within the context of COVID-19 and other zoonotic diseases, could surpass that of drugs, weapons and human trafficking (Doody et al., 2021), made change at a global level imperative and shifted it from the grassroots organisations that had sought change for the animals themselves. Re-positioning the status of animals valued for what they can provide humans, be it coffee or bile, is also underway but without the impetus of human fear and panic over zoonoses it may be slower.

The continued use of animals in research is not so clear cut as debates are played out against the traditional perception of human supremacy. Potential animal-free alternatives, including methods using human cells and tissues (*in vitro* methods), advanced computer-modelling techniques (*in silico* methods) and studies with human volunteers continue to be developed, but must also be embraced by researchers to enact change (PETA, 2023).

REFERENCES

Animals Asia. (2018, April 17). *How a bear called Hong has saved over 600 lives*. https://www.animalsasia.org/intl/social/jills-blog/2018/04/17/how-a-bear-called-hong-has-saved-over-600-lives/

Animals Asia. (2020). *About us*. https://www.animalsasia.org/au/about-us/

Animals Asia. (2021, November 1). *Five things you need to know about bear bile farming*. https://www.animalsasia.org/au/media/news/news-archive/five-things-you-need-to-know-about-bear-bile-farming.html

Baines, J. (2022, February 2). Don't treat animals as spare parts for people. *New Scientist* (3372). https://www.newscientist.com/letter/mg25333720-300-dont-treat-animals-as-spare-parts-for-people-1/

Bekoff, M. (2001). The evolution of animal play, emotions, and social morality: On science, theology, spirituality, personhood, and love. *Zygon, 36*(4), 615–655. https://doi.org/10.1111/0591-2385.00388

Bekoff, M. (2007). *The emotional lives of animals: A leading scientist explores animal joy, sorrow, and empathy–and why they matter*. New World Library.

Challenger, M. (2021). *How to be animal. A new history of what it means to be human*. Canongate Books Ltd.

Córdoba-Aguilar, A., Ibarra-Cerdeña, C.N., Castro-Arellano, I. & Suzan, G. (2021). Tackling zoonoses in a crowded world: Lessons to be learned from the COVID-19 pandemic. *Acta Tropica, 214*, 105780. https://doi.org/10.1016/j.actatropica.2020.105780

Doody, J.S., Reid, J.A., Bilali, K., Diaz, J. & Mattheus, N. (2021). In the post-COVID-19 era, is the illegal wildlife trade the most serious form of trafficking? *Crime Sci, 10*(19), 1–12. https://doi.org/10.1186/s40163-021-00154-9

Galindo-González, J. (2022). Live animal markets: Identifying the origins of emerging infectious diseases. *Current Opinion in Environmental Science & Health, 25*, 100310. https://doi.org/10.1016/j.coesh.2021.100310

Hooper, J. (2022) Contamination: The case of civets, companionship, COVID, and SARS. *Journal of Applied Animal Welfare Science, 25*(2), 167–179. https://doi.org/10.1080/10888705.2022.2028627

Johnson, L.S.M. (2022). Existing ethical tensions in xenotransplantation. *Cambridge Quarterly of Healthcare Ethics, 31*(3), 355–367. https://doi.org/10.1017/S0963180121001055

Liberty Foundation Australia. (2022). *Our 2021-2022 annual update*. https://www.libertyfoundation.org.au/rehoming/our-2021-22-annual-update/

Linzey, A. & Linzey, C. (2018). *The ethical case against animal experiments*. University of Illinois Press. https://www.jstor.org/stable/10.5406/j.ctt2050vt5

Linzey, A. (Ed.), Linzey, C. (Ed.) & Peggs, K. (2015). *Normalising the unthinkable: The ethics of using animals in research*. Oxford Centre for Animal Ethics.

Lu, T., Yang, B., Wang, R. & Qin, C. (2020). Xenotransplantation: Current status in preclinical research. *Frontiers in Immunology, 10*(3060). https://doi.org/10.3389/fimmu.2019.03060

National Health and Medical Research Council (2013). *Australian code for the care and use of animals for scientific purposes*, 8th edition. National Health and Medical Research Council.

PETA. (2023). *Alternatives to animal testing*. In Vitro Methods and More Animal Testing Alternatives | PETA. https://www.peta.org/issues/animals-used-for-experimentation/alternatives-animal-testing/#:~:text=These%20alternatives%20to%20animal%20testing,and%20studies%20with%20human%20volunteers

Sexton, R., Nguyen, T. & Roberts, D.L. (2021). The use and prescription of pangolin in traditional Vietnamese medicine. *Tropical Conservation Science, 14,* 1–13. https://doi.org/10.1177/1940082920985755

Watts, J. (2004). China culls wild animals to prevent new SARS threat. *The Lancet, 363,* 134.

Wikramanayake, E., Pfeiffer, D.U., Magouras, I., Conan, A., Ziegler S., Bonebrake, T.C. & Olson, D. (2021). A tool for rapid assessment of wildlife markets in the Asia-Pacific Region for risk of future zoonotic disease outbreaks. *One Health, 13,* 100279. https://doi.org/10.1016/j.one-hlt.2021.100279

Wildlife Conservation Society (2020). WCS calls for closing live animal markets that trade in wildlife in wake of Coronavirus outbreak. *WCS Newsroom.* https://newsroom.wcs.org/News-Releases/articleType/ArticleView/articleId/13738/WCS-Calls-for-Closing-Live-Animal-Markets-that-Trade-in-Wildlife-in-Wake-of-Wuhan-Coronavirus-Outbreak.aspx

World Animal Protection. (2020). *Beyond wet markets: The many problems with wildlife trade.* https://www.worldanimalprotection.org/blogs/beyond-wet-markets-many-problems-global-wildlife-trade

Wyatt, T., Maher, J., Allen, D., Clarke, N. & Rook, D. (2022). The welfare of wildlife: An interdisciplinary analysis of harm in the legal and illegal wildlife trades and possible ways forward. *Crime, Law and Social Change, 77,* 69–89. https://doi.org/10.1007/s10611-021-09984-9

13

COMPANION ANIMALS AND EMOTION

The previous chapters detailed some confronting examples of mistreatment of animals with limited consideration of their moral standing. This chapter focuses on the more familiar and questions the implications when animals capable of experiencing their own emotional lives are perceived as beloved, but functional commodities for human benefit. This ranges from the companion animal adopted as a buffer to loneliness, to animal-assisted therapy (AAT) drawing on the empathic connection between human and animal, to the implications of factory-farmed domestic dogs and cats bred to meet human demand. Considerable anecdotal and research evidence supports the strength of the human–animal bond to the mutual benefit of both parties, but this bond can also be subject to the complexity and contradictions inherent in any relationship that humans share with animals.

CASE STUDY – WHEN LOVE HURTS

Michelle and Mika huddled together on the cold pavement. Mika wore a dirty jacket that enveloped her slight body. Michelle's pinched face was drained of colour and there was a tremor in the hand resting lightly on Mika's flank. "I should be in hospital right now but I can't leave Mika", Michelle explained as we both gazed at the scruffy terrier curled in a tight ball on Michelle's lap so no part of the dog's body touched the cold, hard concrete. I squatted down to eye level, shivering at the chill that seeped up from the ground and the bitter wind gusting between the high-rise buildings. I tried not to think what it would be like sleeping here.

For 12 months I passed Michelle and Mika camped out on the same city street every day. After the first month, I started bringing dog food for the diminutive terrier and was always delighted to see Michelle's face light up far more than for the coins that infrequently dropped into her broken plastic container. We would talk as Mika ate, then Michelle would painfully rise to take the dog to relieve herself. As the two old friends slowly ambled down the city street to a small oasis of green, the mutual respect within their relationship was evident.

That was 2020, and a pandemic was about to close the city streets and empty the offices of the city workers who dropped coins in Michelle's container. I never saw the two old friends again after the city closed down during the first of several lockdowns, but I like to think they are still together.

DOI: 10.1201/9781003298489-14

Years earlier, as an enthusiastic mental health worker fresh from training, I met Tim and Brutus the Rottweiller. Tim's mental health fluctuated dramatically over the months we worked together, as did his capacity to care for the dog he claimed to love. One 40 degree (Celsius) summer day, Tim tied Brutus outside a hotel and forgot to come back for him. The confused young dog was muzzled and delivered to a small enclosure in a council pound where he paced relentlessly and growled at anyone who approached. Tim swore he could not live without Brutus, but the question had to be asked—would Brutus be better off without Tim? The massive young un-desexed dog covered in flea bite-induced allergies and riddled with intestinal worms may have alleviated his human's loneliness and no doubt would continue to do so willingly for the rest of his life. However, the one-sided anthropocentrism of this relationship could not be ignored. Brutus was a much-loved crutch rather than a sentient creature with feelings and needs of his own. Tim chose not to collect Brutus from the pound, instead leaving it to myself and a not-for-profit rescue group to save Brutus from potential euthanasia because of his aggressive behaviours and chronic skin condition. We arranged desexing, cleared his skin infections, fleas and worms, and eventually the rehabilitated gentle dog was rehomed.

The term human–animal bond describes the connection between humans and animals and encompasses many of the expectations within the relationship. Partially based on the assumption that humans have an innate need to connect with the natural world, animals provide one conduit through which to achieve this in a relationship that ideally is reciprocal. The human–animal bond can thus be defined as "a mutually beneficial and dynamic relationship between people and animals that positively benefits the health and wellbeing of both" (Fine & Griffin, 2022, p. 10). However, the bond may be severed when human expectations are not met, or their life circumstances change and the reciprocity of the bond is no longer mutually beneficial.

Decades of anecdotes and research literature suggest that companion animals can alleviate human loneliness and provide the unconditional love that many people crave. Companion animals offer opportunities for nurturance and social connection to wider society. While Michelle and Mika existed in a mutually beneficial relationship where each provided emotional sustenance to the other, Tim and Brutus shared a different relationship. Tim, through no fault of his own, could neither care for Brutus, nor recognise him as a sentient being who was suffering physically and emotionally.

Companion animals, especially in many Western nations, exemplify the contradictions inherent in the human–animal relationship. Many companion animals are considered beloved family members and treated accordingly. They live a life of luxury with considerable money lavished on feeding, grooming, accessorising and extending their longevity. Other scenarios see humans choosing homelessness rather than accepting housing that excludes their beloved animal companion or risking their own safety to rescue and remain with their animal during natural disasters or war situations. Sociologist Leslie Irvine's thought-provoking book *My Dog Always Eats First: Homeless People and Their Animals* (2015) provides stories of devotion between humans and animals, while anecdotes from natural disaster and war zones further exemplify the bond. There is considerable literature attesting to the strength of the human–animal bond and the physical, emotional and social benefits for the human in this relationship. The attachment that some humans feel for their companion animals has similar neurochemical and hormonal qualities to that of mother–infant interactions (Peralta, 2021), bringing with it a sense of contentment, stability and facilitating a safe environment for curiosity, play and growth. This chemically induced emotional intensity prompts a person to perceive their companion animal's unique personality, emotional capacity and inclusion in the kinship network as a family member or close friend. The companion animal achieves personhood in the eyes of their human despite lacking human language and autonomy.

At the other end of the spectrum, companion animals can also represent commodification of a relationship that has generated a multi-million dollar industry of animal care products and services (DeMello, 2012). This commodification has the potential to result in some animals being abandoned at shelters or euthanised if changes in the human's life circumstances exclude the animal. The popular perception that companion animals can provide a buffer to human anxiety, depression and loneliness raises questions about the long-term welfare of an animal when they are no longer required for their primary function. The moral implications of creating emotional bonds with an animal for practical reasons became evident as the 2020 COVID-19 pandemic reduced human socialisation in many countries.

A defining feature of the COVID-19 pandemic was the sense of isolation that many humans experienced in the wake of this global emergency. Some humans turned to their existing companion animals for support, while others emptied the animal rescue shelters in search of a buffer to the shocking events unfolding around them. A multitude of studies confirmed the protective effect that these animals provided against extreme psychological distress and social isolation for some humans (see, for example, Kogan et al., 2021), providing a sense of purpose, meaning, routine and an overall positive impact on their human's mental and physical functioning. Other studies indicated that many adults viewed their companion animals as a form of supplementary social support, with the added bonus of offering stability, predictability and real-time interaction to counter the technology-based virtual world that dominated human interactions during the pandemic (Jalongo, 2021). There was, however, a darker side to this development as it became evident that a companion animal's quality of life could be influenced by the behaviour and negative emotions of their humans during this period (Shoesmith et al., 2021). Reports of animals being perceived as an added burden of responsibility during an already stressful time, thus negatively impacting the human's quality of life, raised concerns about the reciprocal effect on the animal's emotional wellbeing. Accessing veterinary care and animal food as supermarket shelves emptied and resources and veterinary clinics closed or imposed restrictions was compounded by animal care expenses in a time of income insecurity (Wells et al., 2022).

There is no denying that for some people, companion animals can provide a buffer to loneliness. During one of our street conversations prior to the pandemic, Michelle spoke of the long, lonely nights sleeping on the streets and the tangible support provided by the diminutive Mika. "Mika reminds me that I'm someone, she sees me, she cuts through the absolute aloneness I feel", Michelle explained. "Everyone has to be important to someone and I'm important to Mika and she's important to me".

Even before COVID-19 and the resultant social isolation, companion animals in Korea were perceived as a safeguard against loneliness among a rapidly ageing population and one-person households (Lee et al., 2022). In line with the global increase in demand for adoptable animals, an annual survey of Korean households by the Ministry of Agriculture, Food and Rural Affairs reported a 7% increase in the number of households including at least one companion animal in 2020. Further research indicated companion dogs, in particular, were reported to reduce negative feelings about the pandemic, while simultaneously maintaining human physical health by increasing mobility around neighbourhoods when permitted. With dog walking being one of the few reasons for leaving the house in many countries, animal shelters emptied rapidly.

In many cases, human–animal bonds increased when mutually experiencing the stress and hardships associated with the pandemic although China presented a picture of contrasts during this time. Reports of abandonment or killing of companion animals emerged as some members of the community were unable to separate companion animals from animals in wildlife markets. These reports were juxtaposed against poignant images of the Wuhan Small Animal Protection Association, a non-government organisation, entering the empty homes of Wuhan residents unable to return to their animals when locked out of the city. This ethical duty of care to vulnerable animals in Wuhan, the perceived epicentre of the pandemic, was conducted against a backdrop of global panic and blame (Yin et al., 2020).

Chinese communities were not unique in their initial uncertainty about cats and dogs contracting COVID and transmitting it to humans. Globally, there were some anecdotal cases of panic abandonment, relinquishment and euthanasia of companion animals but this soon settled when fears proved unfounded. As country after country went into lockdown, adoptions increased in an attempt to ease feelings of isolation or to bring pleasure to children, sometimes with minimal consideration of the long-term implications of adopting sentient creatures with emotions, personalities and social needs of their own. It quickly became evident that there were consequences to confining human and non-human animals in an unnatural pressure cooker of emotions. Dogs spend up to half their day sleeping, cats even more. When unable to escape for downtime in a bustling household, the potential for problematic interactions escalated. Behavioural issues, anxiety, fearfulness, lack of external socialisation opportunities or lack of downtime to buffer negative effects raised concerns about the viability of some relationships for both humans and animals (Bowen et al., 2021).

Research focus gradually shifted to examine the impact on animals locked up with humans and potentially having to deal with the human's increased emotional needs. Cross-species emotional contagion whereby the perception of an affective behavioural change in one party can automatically trigger a response in the other party has been observed between humans and their companion animals. Anecdotally, people take pride in their companion animal's ability to sense their moods and still provide unconditional positive regard. However, reports continued to emerge about changes in companion animals' behaviours as the pandemic lingered, with no definitive way to determine if these animals were mirroring their humans' state of worry or were the recipients of their humans' projected feelings of stress and anxiety. Where animals become depositories for human's projected feelings, there can be a blurring of boundaries between humans and animals as the animal adapts to the human's needs and the human interprets the animal's behaviours within the context of their own projected feelings. An anxious human will interpret their animal's behavioural responses as anxious, becoming more anxious themselves in an ongoing feedback loop that can leave the animal behaviourally disoriented. Reports from multiple countries cited behavioural changes in companion animals, including an increase in needy, clingy and attention-seeking behaviours; noise; anxiety and nervousness; and some unusual aggressive behaviours

(Esam et al., 2021). A common factor in many reports of these stress-related behaviours was the presence of a human who was not coping well with life in a pandemic.

With the easing of pandemic restrictions and gradual return to pre-pandemic real-time human interactions, a different set of concerns emerged. Dependency on humans has been selectively bred into companion animals, especially dogs. The dependency traits that humans desired and admired in their animals—strong attachment, companionship, empathy—and that had proven so affirming during social isolation became problematic when leaving the house to return to the outside world. These coveted traits were no longer endearing as some animals responded badly to being shelved like commodities when no longer required. Figure 13.1 shows a solitary dog sitting at a window once its human family were no longer confined to the house. Other outcomes included barking and destructive behaviours when no longer receiving adequate human interaction.

Anecdotal reports of animals being relinquished to animal shelters, rehomed, abandoned or even euthanised surfaced with the most common explanation centred around the animal's problematic behaviours, especially aggression. Among dogs, growling, nipping and biting increased, according to the United Kingdom Dog Trust (2020), with one emergency department reporting three times as many dog bites presenting among children in 2020 (Jalongo, 2021). These were potentially blamed on the surge in "pandemic

Figure 13.1 A solitary dog sits at a window and gazes outside. Photograph supplied by the author.

puppies" adopted to counter loneliness and provide physical stimulation, often with unrealistic expectations of what animals needed long-term. Larger dogs seemed at greater risk of relinquishment as individuals and families returned to normal external activities and dogs who had never been alone were required to remain behind. Kittens and cats also faced an uncertain future in some families, sometimes reinforced by erroneous societal stereotypes that cats are more independent and therefore more easily abandoned.

On 31 December 2022, a small post appeared on the Facebook page of "Ingrid's Haven", a no-kill not-for-profit cat shelter in Victoria, Australia. Over more than two decades of animal rescue, Ingrid Arving has saved the lives of thousands of abandoned and unwanted cats, including the FIV (Feline Immunodeficiency Virus) cats who are often overlooked for adoption. With the help of a dedicated team of volunteers, cats from Ingrid's Haven have the opportunity to feel safe again, to experience positive emotions and to express their unique personalities. Ingrid's poignant words succinctly describe the forgotten casualties from a pandemic whose ripple effect continued.

"It is now time to say goodbye to 2022, a year I will not miss one bit. It has been a horrendous year with all the COVID cats being dumped, abandoned, surrendered and having kittens. All rescues that I know of are doing their utmost to help these souls".

Reproduced with permission from Ingrid Arving, Ingrid's Haven.

When contacted about this worrying post, Ingrid expanded: "I said no to a lot of people wanting a cat for the lockdowns and was told how rude I was and unfeeling [...] Lots of 'rescues' and breeders, both backyard and registered, jumped on the bandwagon and I see cats coming back via my pound … It's been such a heartbreak; how do you explain to an animal that had 24/7 attention that it is no longer wanted?"

Applying the Zoological Emotional Scale in Figure 13.2 to the life history of one kitten provides some background as to this phenomenon and identifies points where change could be enacted.

THE CASE OF BOOTS

The children's excitement was palpable. A week earlier they had selected the bed, toys and food for the kitten who had been born from the unexpected pregnancy of a friend's cat. At five weeks old, the tiny ginger tabby kitten, not yet old enough to be away from her mother, joined the lockdown household. At four months old, the kitten was relinquished to an animal shelter. The intervening weeks were a roller coaster of excitement, fear and anger for the three children and two adults as many of their expectations failed to materialise. In a scenario played out across many households, the reality of caring for a young animal with a unique personality and needs of her own proved too difficult to absorb into an already stressed household.

Figure 13.2 demonstrates Boots' transition from a new acquisition based on fallacy, to both an initial friend with emotional capacity to bond with the children and a functional commodity to keep the children entertained, and finally to foe after Boots retaliated to rough treatment and drew blood. Despite well-publicised statements from animal shelters and welfare agencies stressing that animals are not toys to be played with then put away when done, rather they are a lifelong commitment requiring attention, enrichment, love and training, this message is not always heeded, especially for cute, cuddly infant animals with paedomorphic features. This can result in relinquishment, abandonment or death.

1. FALLACY

Boots entered a household that
was inexperienced with animals
and believed that companion
animals and children form an
instant bond, grow up together
and the children become
responsible caregivers.

2. FRIEND

Boots was overwhelmed
when she first arrived, having
just left her mother and
siblings. Her neediness and
compliance positioned her as
a perfect living toy.

3. FUNCTIONAL

Boots provided an instant
diversion to the boredom and
loss of companionship
experienced by children during
COVID lockdowns. Initially the
children complied with the
additional chores.

4. FOE

Boots learned to fear the
children's rough play and
started to scratch, bite and
urinate outside the litter tray.
Boots was locked alone in the
garage then relinquished to an
animal shelter.

Figure 13.2 The Zoological Emotional Scale showing how any attribution of emotional capacity to Boots' transitioned from the fallacy to foe domains.

ANIMAL-ASSISTED INTERVENTIONS

The privilege of connecting with an animal carries with it a responsibility to ensure their basic needs are met and their daily experiences are largely positive. This responsibility is particularly important when animals are enlisted in activities such as human mental health care that specifically benefit humans, but that may potentially carry an emotional risk to the animal's wellbeing (Peralta, 2021). Animal-assisted interventions (AAIs) are structured, goal-directed interventions incorporating animals for human therapeutic gains and improved health and wellness. AAI can range from animal-assisted activities (AAAs) such as hospital visits and stress-reduction visits at universities and workplaces, to AAT such as psychotherapy and counselling, physical therapy, social work and assistance and service dogs (Jones et al., 2018).

The potential benefits of including animals in therapeutic and educational settings for human benefit have attracted considerable discussion since child psychologist Boris Levinson included his dog Jingles into therapy and subsequently coined the term "pet therapy" in the 1960s. Initially fuelled by rich anecdotal data to validate the importance of the animal's role to human wellbeing, ongoing studies faced criticisms related to lack of consistency in sample size, demographics, research design and appropriate controls. These early studies also focused on the perceived benefits to humans, thus omitting the wellbeing of the animal without whom AAI would be impossible. Reported reductions in human stress and anxiety when including an animal in therapeutic, work or educational settings failed to address the animal's experience, especially within the context of the animal's own emotional capacity and the potential for emotional contagion when exposed to negative human emotions.

The next wave of research pushed for some form of regulatory standard of practice to address both the efficacy of the models and ensure safe procedures for the human

(Jones et al., 2018). As it became evident that animals could act as emotional buffers and provide a sense of safety for some humans, attention subsequently turned to potential risks for the animals if not closely monitored (Fine et al., 2019). Research findings that some dogs displayed increased levels of the stress hormone cortisol following therapy visits raised concerns as animals in diverse health and educational settings became mainstream.

While much of the animal welfare research has focused on dogs, one of the more popular AAA animals, a range of other animals including cats, rabbits, rodents, birds, horses, goats, reptiles and fish were enlisted as mechanisms to improve human mental health and wellbeing in a range of settings. For example, observing a fish aquarium has been shown to have relaxing properties, while goats are finding a niche in the therapy world by visiting some Japanese hospitals, assisted living homes, nursing homes and rehabilitation centres.

The inclusion of other animals whose needs and responses were not always as familiar as those of the domestic dog provided further impetus to ensure that the quality of life for all animals was monitored in a way that recognised their unique characteristics. Cross-species behavioural and physiological parameters—including heart rate, cortisol levels and body language—and proper supervision provided some guards against negative impacts on the animals as succinct guidelines for animal handling and close environmental monitoring across a range of settings began to evolve (Peralta, 2021).

Different animals will react differently to AAIs, especially when the AAI involves meeting a wide range of humans, many with mental or physical health problems. Some interactions, including unprovoked attention and rough contact, can be potential stressors for dogs, while any physical contact for some cats can be a significant stressor. Poor horse and rider matching in equine AAIs can result in stress for the horse. Some reptiles can develop tolerance for handling, but there is evidence that they do not crave affection or human contact like some mammals used in AAI and can become defensive when feeling threatened.

Enlisting complex beings with varying degrees of emotional capacity into settings specifically focused on human physical or psychological wellbeing has moral and ethical implications and requires consideration from the animal's perspective, not just human wellbeing. Mellor's Five Domains may provide one lens through which to examine this relationship (Fine & Griffin, 2022). The bond between handler and animal usually provides assurance that the first four domains—that is, (1) freedom from hunger and thirst; (2) freedom from discomfort; (3) freedom from pain, injury or disease; and (4) freedom to express normal behaviour—are satisfactorily met, but Domain 5, freedom from fear and distress, can fluctuate. Domain 5 recognises opportunities leading to a positive affective state, not just the removal of situations resulting in a negative state. The animal's mental status is critical to Domain 5, and it can change over time as the animal grows older or the cumulative burden of emotional contagion becomes too great. Unpredictable, unfamiliar or aversive stimuli in some therapeutic or educational settings can be stressful, as evidenced by the therapy dogs in hospital settings recording increased salivary cortisol levels (Ng et al., 2014). Dogs, in particular, may retain a wish to please, masking that they are struggling physically and emotionally in their role of supporting humans.

Whether companion or AAI animal, there is no doubt that strong attachment bonds can form between humans and animals. In most cases, the animal is perceived as an individual whose moral standing receives concern and respect. Exceptions occur where the animal becomes a burden and the human seeks to divest themselves of the responsibility, but for many people the human–animal bond is enduring. Puppy mills, however, provide an example of where an animal starts and ends their life as a commodity with no moral standing or opportunity to demonstrate positive emotions. Their sole purpose is to produce puppies to be loved by someone else, then to be discarded when they are too old or worn out.

PUPPY FARMS

Many dogs become valued members of a human family, described as surrogate children, siblings and companions, and loved, pampered and grieved when they die. Positioning a companion dog on the Zoological Emotional Scale is relatively easy for their human: the dog is perceived as a friend, capable of loving and being loved, and displaying an assortment of emotions from joy and happiness to sadness and guilt. The companion dog may also be perceived as functional through the provision of unconditional positive regard, emotional support, a buffer to loneliness, motivation to exercise and a sense of safety. Death of some beloved dogs attracts the same rituals as human death, including cremation or burial, and continuing bonds expressions such dreams, fond memories and memorials (for example, see works by Packman et al., 2012).

Companion dogs, especially the hybrid breeds or designer dogs, sometimes share a dark heritage arising from the puppy farm (also known as a puppy factory or puppy mill) in which they were bred. These breeding facilities can operate under inadequate conditions that fail to meet the behavioural, social, emotional and physiological needs of mother or puppies. At their worst, they can be over-crowded and unhygienic with the mothers enduring endless cycles of litters until they are eventually "bred to death" (RSPCA, 2022). In October 2022, the tiny Oscar, survivor of a puppy farm, icon, hero and best friend, died at the age of 16 years. Oscar was rescued after spending five years in a puppy farm as a stud dog. Emaciated and psychologically traumatised, Oscar's emotive face (see Figure 13.3) and resilience inspired Oscar's Law, a movement to get justice for factory-farmed dogs in Australia. Oscar's Law, like Lucy's Law in the United Kingdom and other Anti-Puppy Mill legislation, campaigned relentlessly to shut down the puppy production lines that churned out expensive commodities to be transformed into a human's best friend. The death of this public figure prompted an outpouring of grief and pledges to continue Oscar's work on behalf of the faceless animals still serving human needs (Oscar's Law, 2023).

Despite legislation and grass-root movements, pandemic-prompted demand for dogs during COVID-19 lockdowns saw illicit puppy breeding and online scams increase in some countries. The juxtaposition of offspring as beloved family members to be loved and grieved compared to the mothers and stud dogs as commodities to be "bred to death" can create feelings of guilt for humans who discover that they have unsuspectingly purchased a puppy farm dog. This was evident in the events described to me by a young man who impulsively bought a Labradoodle (Labrador crossed with poodle) from an online advertisement. It was only when the puppy displayed a series of potential congenital health problems that he was forced to face the origins of his impulse-purchase. When his veterinarian asked for more information on the parents' health, the puppy seller was nowhere to be found and the young man went into considerable debt to heal the puppy with whom he had become firmly bonded. Forced to face the potential origins of his beloved dog, he was visibly upset: "It breaks my heart

Figure 13.3 Oscar's emotive face inspired Oscar's Law. Photograph provided by Oscar's Law and reproduced with permission.

to think of his mother like that. But there was nothing I could do by then. I've spent thousands of dollars getting the little fella right, and I wouldn't have it any other way. But his mother haunts me. I feel so stupid knowing I've given money to her abusers".

Comparing the beloved Labradoodle with the dog's mother on the Zoological Emotional Scale shows a clear divergence despite there being no physical or neurological differences between the two genetically connected dogs. Their positioning and the way their lives and deaths would unfold are based on the needs of humans as the more dominant species. Figure 13.4 contrasts the subjective perceptions of the two dogs as friend or functional based on the privileging of one animal over the other.

IMPLICATIONS

Humans are animals, despite distancing themselves from animal-like actions and claiming supremacy. There remains embedded in the human psyche an innate need to re-connect with nature, even as the natural world is irrevocably damaged by the collective actions of humanity. For urban dwellers, companion animals can provide a conduit to the natural world, enhanced by a positive emotional experience and avoidance of negative feelings such as loneliness and worthlessness. Attachment to a companion animal, however, invokes responsibilities that can be overlooked through lack of awareness or disregarded to avoid cognitive dissonance.

Like domestic-farmed animals, some companion animals have been selectively bred to meet human needs resulting in a dependency that is not always to their benefit. Heartbreaking images of companion animals forcibly abandoned during war events and cared for by a handful of humans at great risk demonstrate how fragile this dependency can become when humans must leave. Of equal concern is the intentional or inadvertent

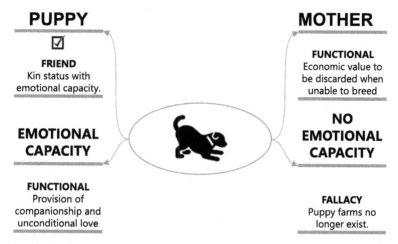

PUPPY

☑
FRIEND
Kin status with
emotional capacity.

**EMOTIONAL
CAPACITY**

FUNCTIONAL
Provision of
companionship and
unconditional love

MOTHER

FUNCTIONAL
Economic value to
be discarded when
unable to breed

**NO
EMOTIONAL
CAPACITY**

FALLACY
Puppy farms no
longer exist.

Figure 13.4 The Zoological Emotional Scale compares the life of a puppy farm mother and her son.

exploitation of this dependency to the detriment of the companion animal's emotional wellbeing. In a real-life parody of Universal Picture's 2016 animated movie *The Secret Life of Pets*, some animals obtained during global lockdowns to allay human isolation did not adjust when their humans once again re-entered the human world. When left to emotionally fend for themselves, the behavioural problems of these animals were more likely to result in abandonment than the high adventure and self-entertainment depicted in this movie that subtly reinforced social norms around domestic animals as property.

Potential inadvertent emotional distress within a supportive domestic or AAI relationship is neither easily identified nor easily rectified, partly because of the pervasive perceptions of human ownership and dominance. Companion animals do not have personhood and are therefore the property of the human, albeit property kept in a comfortable and supportive environment. It is too late to turn back the evolutionary changes that have created such dependency in companion animals, but it is not too late to recognise them as emotional beings capable of living meaningful lives in their own right, not just as adjuncts to the human experience. This requires a change in perception and reframing of the question: "What can this animal do for me?" to "What can we do together?"

REFERENCES

Bowen, J., García, E., Darder, P., Argüelles, J. & Fatjó, J. (2021). The effects of the Spanish COVID-19 lockdown on people, their pets, and the human-animal bond. *Journal of Veterinary Behavior, 40*, 75–91. https://doi.org/10.1016/j.jveb.2020.05.013

DeMello, M. (2012). *Animals and society: An introduction to human-animal studies.* Columbia University Press.

Esam, F., Forrest, R. & Waran, N. (2021). Locking down the impact of New Zealand's COVID-19 alert level changes on pets. *Animals, 11*(3), 758. https://doi.org/10.3390/ani11030758

Fine, A.H., Beck, A.M. & Ng, Z. (2019). The state of animal-assisted interventions: Addressing the contemporary issues that will shape the future. *International Journal of Environmental Research and Public Health, 16*(20), 3997. https://doi.org/10.3390/ijerph16203997

Fine, A.H. & Griffin, T.C. (2022). Protecting animal welfare in animal-assisted intervention: Our ethical obligation. *Seminars in Speech and Language, 43*(1), 8–23. https://doi.org/10.1055/s-0041-1742099

Irvine, L. (2015). *My dog always eats first: Homeless people and their animals.* Lynne Rienner Publishers.

Jalongo, M.R. (2021). Pet keeping in the time of COVID-19: The canine and feline companions of young children. *Early Childhood Education Journal, 18*, 1–11. https://doi.org/10.1007/s10643-021-01251-9

Jones, M.G., Rice, S.M. & Cotton, S.M. (2018). Who let the dogs out? Therapy dogs in clinical practice. *Australasian Psychiatry, 26*(2), 196 -199. https://doi.org/10.1177/1039856217749056

Kogan, L.R., Currin-McCulloch, J., Bussolari, C., Packman, W. & Erdman, P. (2021). The psychosocial influence of companion animals on positive and negative affect during the COVID-19 pandemic. *Animals, 11*(7), 2084. https://doi.org/10.3390/ani11072084

Lee, H.-S., Song, J.-G. & Lee, J.-Y. (2022). Influences of dog attachment and dog walking on reducing loneliness during the COVID-19 pandemic in Korea. *Animals, 12*, 483. https://doi.org/10.3390/ani12040483

Ng, Z.Y., Pierce, B.J., Otto, C.M., Buechner-Maxwell, V.A., Siracusa, C. & Were, S.R. (2014). The effect of dog-human interaction on cortisol and behavior in registered animal-assisted activity dogs. *Applied Animal Behavior Science, 159*, 69–81. https://doi.org/10.1016/j.applanim.2014.07.009

Oscar's Law (2023). *Who is Oscar?* https://www.oscarslaw.org/who-is-oscar.htm

Packman, W., Carmack, B.J. & Ronen, R. (2012). Therapeutic implications of continuing bonds expressions following the death of a pet. *OMEGA - Journal of Death and Dying, 64*(4), 335–356. https://doi.org/10.2190/OM.64.4.d

Peralta, J.M. (2021). Animals' perspective and its impact on welfare during animal-assisted interventions. In J. M. Peralta & A. H. Fine (Eds.), *The welfare of animals in animal-assisted interventions* (pp. 1–20). Springer. https://doi.org/10.1007/978-3-030-69587-3_1

RSPCA, (2022). *Puppy factories.* https://rspcavic.org/learn/puppy-factories/?ReferredFrom=adwd-n&gclid=Cj0KCQjwlK-WBhDjARIsAO2sErSfIkvAogf1gFOh7h-EtniFY8VsvRtdqlv_minOD-KJ1nLqAPX3NxbEaAmT_EALw_wcB

Shoesmith, E., Santos de Assis, L,Shahab, L., Ratschen, E., Toner, P., Kale, D., Reeve, C. & Mills, D.S. (2021). The perceived impact of the first UK COVID-19 lockdown on companion animal welfare and behaviour: A mixed-method study of associations with owner mental health. *International Journal of Environmental Research and Public Health, 18*, 6171. https://doi.org/10.3390/ijerph18116171

Wells, D.L., Clements, M.A., Elliott, L.J. & Meehan, E.S., Montgomery, C.J. & Williams, G.A. (2022). Quality of the human-animal bond and mental wellbeing during a COVID-19 lockdown. *Anthrozoös, 35*(6), 847–866. https://doi.org/10.1080/08927936.2022.2051935

Yin, D., Gao, Q., Zhu, H. & Li, J. (2020). Public perception of urban companion animals during the COVID-19 outbreak in China. *Health Place, 65*, 102399. https://doi.org/10.1016/j.healthplace.2020.102399

CONCLUSION: THE NEED TO LISTEN

14

Sometimes, it is easy to forget that we share this planet with the most wonderful collection of unique beings. For many people, animals are on the periphery of their existence, perhaps represented by a companion animal, a trip to the zoo, nature documentaries and the insects that encroach on domestic spaces. Connecting the death of animals to the plastic-wrapped meat in supermarkets, milk cartons, square blocks of cheese, leather shoes and medicines maintaining human health and longevity is a harder concept to grasp. To do this requires delving into the cognitive dissonance and psychic numbing that has allowed human domination of animals and the natural world to the detriment of the planet. Acknowledging that these animals have positive and negative feelings is an even harder task on which to embark as that requires acknowledging the human-inflicted separation, discrimination and suffering of individual animals, groups and whole species.

Recently, I watched some ants working in unison to shift the carcass of a dead earthworm. As my shadow stretched over them, they paused in their work as if checking the safety of continuing this mammoth task. I will never know what they were collectively thinking, but I was able to hazard a guess as to the possible uncertainty that temporarily halted their group actions. My dog took advantage of my momentary stillness to lean contentedly against my legs and I easily interpreted his actions as affection and happiness to be out walking with me. Behavioural indicators of animal perception, feelings and emotions are to be found in the mundane, open to all of us to observe and take note. Some are easier than others to interpret or guess, often based on visual appearance or familiarity, but that does not mean the unfamiliar should be disregarded. Every animal has their own life and story, and as humans, the onus is on us to listen. Some stories and feelings are harder to hear or interpret with our limited human capacities, but that does not mean they do not exist.

There are so many complex and interwoven factors that can interrupt a person's perceptions of animal emotional abilities, ranging from cultural and spiritual beliefs to the disgust factor for animals that do not resemble us. One by one these are being overturned, but not fast enough for some of the unique species with whom we share this planet. There is now no doubt that vertebrate and invertebrate animals can have positive and negative feelings and experience pain. This will never be identical to the human experience but it is a valid reality in the lives of each animal. Humans dominate the world, meaning the responsibility is on us to ensure social norms, laws and policies

DOI: 10.1201/9781003298489-15

are disengaged from the assumptions and stereotypes inherent in a flawed hierarchy of animal abilities. Every animal has a place in their natural environment, making them all functional members of the world if given the chance.

Despite the growth in scientific research supporting the inner lives of animals and their emotional capacity, there are still barriers to disassemble before humanity embraces the concept that they do not hold a monopoly on emotion, spirituality and morality. While it was not possible to consider every animal species in this book, the Zoological Emotional Scale is provided as a framework to give voice, albeit still a human voice, to the emotional world of animals. Each person will have differing and at times conflicting perceptions of animals, their rights and their place in a human-dominated society, often influenced by their role in relation to animals and animal welfare, but understanding the basis of beliefs and attitudes is an important step in activating the cycle of change.

For the animals and situations omitted from these pages, I apologise. The dark shadow of human control has encroached on so many aspects of animal wellbeing that it was impossible to address every scenario where change is required. We are living in a time of animal emergency, whether they are tiny insects or massive whales, and it is all because of human activities. Change is needed, but understanding how to change is never easy. Humans have already taken so much from animals to the point where the survival of many animals is now dependent on human actions. It is time to give back by recognising and responding to their needs as sentient creatures with the capacity for emotional lives of their own. They may not be emotional lives as we know them, but that is irrelevant.

Look for emotion in the day-to-day and if you cannot see it, it does not mean it is not there. It just means it is different to your experience and the experience of other species. A bee who appears perplexed at the loss of a favourite flower-bed is no different to the magpie whose long-term nesting site was destroyed in the name of human development or the wolf who strays on to forbidden human territory and is killed. You do not need to read their mind to recognise that human actions have affected each one in a bee-, magpie- or wolf-specific way. As animals increasingly lose their habitats, the planet will ultimately lose the animals, but what threatens the animals also threatens humanity.

Cognitive dissonance is unpleasant but can no longer be ignored. Animal products come from animals who are aware and suffering in some very distressing environments. The moral dilemma of having to choose one species to live and one to die is difficult. Humans have changed ecosystems through the introduction of non-native species or the destruction of native habitat. Should the feral cat be allowed to decimate native animals when it is surviving in an environment to which it did not ask to join? Should native kangaroos be culled for intruding on the agricultural land that was once theirs? This book has not set out to answer these big questions that have no right or wrong answer when animals will die whatever is decided. Rather, the aim is to reject the hierarchical systems of categorisation that position humans, as creators of the hierarchy, at the top. These false hierarchies have contributed towards an underlying bias towards empathy for the familiar animals that can more easily be related to and a tendency to ignore or downgrade those on the lower levels.

Instead, this book seeks to encourage people to start from their own point of familiarity with the animal and consider the important questions: "Why do I think this way and are there other ways of perceiving this animal?" While this may create feelings of cognitive dissonance, it can also start the process of change to address negative stereotypes and outcomes for that animal.

Dedicated humans are doing their best to preserve endangered species through zoos and captive breeding programmes, but long-term, if the animals' ecosystem has collapsed or been degraded so as to be uninhabitable for the captive-bred animal, the problem remains. Does having the only samples of a species compensate for the global loss of biodiversity? That question is only one of many in our relationship with animals. Another question that can no longer be ignored is the impact of zoonoses as humans increasingly encroach on animal land or confine them in crowded unnatural situations. The increase in zoonoses has seen some change based on the need to protect human lives and economies. If the animals benefit, that is a bonus even if action is taken for a human-centric reason. Change is also happening in the growing social media platforms of animal rescue and sanctuaries. This is happening one animal at a time, but it is happening.

I am still no closer to resolving the murky middle of animal research, but I have taken a stance on eating and using animal products. My house is now a refuge for the old and discarded animals of society where there is no "them and us" dichotomy. Change can be slow, but humans have accelerated it in ways that were once unimaginable. It therefore rests on humans to reverse the change with a shift in inherited beliefs and attitudes and revisiting the individual contributions to be made. This is a book about the meaning of animal emotion, and it should never be forgotten that humans are animals. Enlisting human animal emotion by breaking down the protective barriers of cognitive dissonance, psychic numbing and activating the cycle of change is an important step in sharing Earth with all the animals who may be at risk or suffering.

INDEX

Printed in the United States
by Baker & Taylor Publisher Services